Memory Architecture Exploration
for Programmable Embedded Systems

MEMORY ARCHITECTURE
EXPLORATION FOR PROGRAMMABLE
EMBEDDED SYSTEMS

PETER GRUN
Center for Embedded Computer Systems,
University of California, Irvine

NIKIL DUTT
Center for Embedded Computer Systems,
University of California, Irvine

ALEX NICOLAU
Center for Embedded Computer Systems,
University of California, Irvine

Kluwer Academic Publishers
Boston/Dordrecht/London

Distributors for North, Central and South America:
Kluwer Academic Publishers
101 Philip Drive
Assinippi Park
Norwell, Massachusetts 02061 USA
Telephone (781) 871-6600
Fax (781) 681-9045
E-Mail: kluwer@wkap.com

Distributors for all other countries:
Kluwer Academic Publishers Group
Post Office Box 322
3300 AH Dordrecht, THE NETHERLANDS
Telephone 31 786 576 000
Fax 31 786 576 254
E-Mail: services@wkap.nl

 Electronic Services < http://www.wkap.nl>

Peter Grun, Nikil Dutt and Alex Nicolau
Memory Architecture Exploration for Programmable Embedded Systems

ISBN 978-1-4419-5329-2   e-ISBN 978-0-306-48095-9
Printed on acid-free paper.

Printed in the United States of America.

Contents

List of Figures

List of Tables

Preface

Continuing advances in chip technology, such as the ability to place more transistors on the same die (together with increased operating speeds) have opened new opportunities in embedded applications, breaking new ground in the domains of communication, multimedia, networking and entertainment. New consumer products, together with increased time-to-market pressures have created the need for rapid exploration tools to evaluate candidate architectures for System-On-Chip (SOC) solutions. Such tools will facilitate the introduction of new products customized for the market and reduce the time-to-market for such products.

While the cost of embedded systems was traditionally dominated by the circuit production costs, the burden has continuously shifted towards the design process, requiring a better design process, and faster turn-around time. In the context of programmable embedded systems, designers critically need the ability to explore rapidly the mapping of target applications to the complete system. Moreover, in today's embedded applications, memory represents a major bottleneck in terms of power, performance, and cost.

The near-exponential growth in processor speeds, coupled with the slower growth in memory speeds continues to exacerbate the traditional processor-memory gap. As a result, the memory subsystem is rapidly becoming the major bottleneck in optimizing the overall system behavior in the design of next generation embedded systems. In order to match the cost, performance, and power goals, all within the desired time-to-market window, a critical aspect is the Design Space Exploration of the memory subsystem, considering all three elements of the embedded memory system: the application, the memory architecture, and the compiler early during the design process.

This book presents such an approach, where we perform Hardware/Software Memory Design Space Exploration considering the memory access patterns in the application, the Processor-Memory Architecture as well as a memory-aware compiler to significantly improve the memory system behavior. By exploring a design space much wider than traditionally considered, it is possible to generate substantial performance improvements, for varied cost and power footprints.

In particular, this book addresses efficient exploration of alternative memory architectures, assisted by a "compiler-in-the-loop" that allows effective matching of the target application to the processor-memory architecture. This new approach for memory architecture exploration replaces the traditional black-box view of the memory system and allows for aggressive co-optimization of the programmable processor together with a customized memory system.

The book concludes with a set of experiments demonstrating the utility of our exploration approach. We perform architecture and compiler exploration for a set of large, real-life benchmarks, uncovering promising memory configurations from different perspectives, such as cost, performance and power. Moreover, we compare our Design Space Exploration heuristic with a brute force full simulation of the design space, to verify that our heuristic successfully follows a true pareto-like curve. Such an early exploration methodology can be used directly by design architects to quickly evaluate different design alternatives, and make confident design decisions based on quantitative figures.

Audience
This book is designed for different groups in the embedded systems-on-chip arena.

First, the book is designed for researchers and graduate students interested in memory architecture exploration in the context of compiler-in-the-loop exploration for programmable embedded systems-on-chip.

Second, the book is intended for embedded system designers who are interested in an early exploration methodology, where they can rapidly evaluate different design alternatives, and customize the architecture using system-level IP blocks, such as processor cores and memories.

Third, the book can be used by CAD developers who wish to migrate from a hardware synthesis target to embedded systems containing processor cores and significant software components. CAD tool developers will be able to review basic concepts in memory architectures with relation to automatic compiler/simulator software toolkit retargeting.

Finally, since the book presents a methodology for exploring and optimizing the memory configurations for embedded systems, it is intended for managers and system designers who may be interested in the emerging embedded system design methodologies for memory-intensive applications.

Acknowledgments

We would like to acknowledge and thank Ashok Halambi, Prabhat Mishra, Srikanth Srinivasan, Partha Biswas, Aviral Shrivastava, Radu Cornea and Nick Savoiu, for their contributions to the EXPRESSION project.

We thank the funding agencies who funded this work, including NSF, DARPA and Motorola Corporation.

We would like to extend our special thanks to Professor Florin Balasa from the University of Illinois, Chicago, for his contribution to the Memory Estimation work, presented in Chapter 2.

We would like to thank Professor Kiyoung Choi and Professor Tony Givargis for their constructive comments on the work.

Chapter 1

INTRODUCTION

1.1 Motivation

Recent advances in chip technology, such as the ability to place more transistors on the same die (together with increased operating speeds) have opened new opportunities in embedded applications, breaking new ground in the domains of communication, multimedia, networking and entertainment. However, these trends have also led to further increase in design complexity, generating tremendous time-to-market pressures. While the cost of embedded systems was traditionally dominated by the circuit production costs, the burden has continuously shifted towards the design process, requiring a better design process, and faster turn-around time. In the context of programmable embedded systems, designers critically need the ability to explore rapidly the mapping of target applications to the *complete* system. Moreover, in today's embedded applications, memory represents a major bottleneck in terms of power, performance, and cost [Prz97]. According to Moore's law, processor performance increases on the average by 60% annually; however, memory performance increases by roughly 10% annually. With the increase of processor speeds, the processor-memory gap is thus further exacerbated [Sem98].

As a result, the memory system is rapidly becoming the major bottleneck in optimizing the overall system behavior. In order to match the cost, performance, and power goals in the targeted time-to-market, a critical aspect is the Design Space Exploration of the memory subsystem, considering all three elements of the embedded memory system: the application, the memory architecture, and the compiler early during the design process. This book presents such an approach, where we perform Hardware/Software Memory Design Space Exploration considering the memory access patterns in the application, the Processor-Memory Architecture as well as a memory-aware compiler, to significantly improve the memory system behavior. By exploring a design space much wider than traditionally considered, it is possible to generate substantial performance improvements, *for varied cost and power footprints.*

1.2   Memory Architecture Exploration for Embedded Systems

Traditionally, while the design of programmable embedded systems has focused on extensive customization of the processor to match the application, the memory subsystem has been considered as a black box, relying mainly on technological advances (e.g., faster DRAMs, SRAMs), or simple cache hierarchies (one or more levels of cache) to improve power and/or performance. However, the memory system presents tremendous opportunities for hardware (memory architecture) and software (compiler and application) customization, since there is a substantial interaction between the application access patterns, the memory architecture, and the compiler optimizations. Moreover, while real-life applications contain a large number of memory references to a diverse set of data structures, a significant percentage of all memory accesses in the application are often generated from a few instructions in the code. For instance, in Vocoder, a GSM voice coding application with 15,000 lines of code, 62% of all memory accesses are generated by only 15 instructions. Furthermore, these instructions often exhibit well-known, predictable access patterns, providing an opportunity for customization of the memory architecture to match the requirements of these access patterns.

For general purpose systems, where many applications are targeted, the designer needs to optimize for the average case. However, for embedded systems the application is known apriori, and the designer needs to customize the system for this specific application. Moreover, a well-matched embedded memory architecture is highly dependent on the application characteristics. While designers have traditionally relied mainly on cache-based architectures, this is only one of many design choices. For instance, a stream-buffer may significantly improve the system behavior for applications that exhibit stream-based accesses. Similarly, the use of linked-list buffers for linked-lists, or SRAMs for small tables of coefficients, may further improve the system. However, it is not trivial to determine the most promising memory architecture matched for the target application.

Traditionally, designers begin the design flow by evaluating different architectural configurations in an ad-hoc manner, based on intuition and experience. After fixing the architecture, and a compiler development phase lasting at least an additional several months, the initial evaluation of the application could be performed. Based on the performance/power figures reported at this stage, the designer has the opportunity to improve the system behavior, by changing the architecture to better fit the application, or by changing the compiler to better account for the architectural features of the system. However, in this iterative design flow, such changes are very time-consuming. A complete design flow iteration may require months.

Alternatively, designers have skipped the compiler development phase, evaluating the architecture using hand-written assembly code, or an existing compiler for a similar Instruction Set Architecture (ISA), assuming that a processor-specific compiler will be available at tape-out. However, this may not generate true performance measures, since the impact of the compiler and the actual application implementation on the system behavior may be significant. In a design space exploration context, for a modern complex system it is virtually impossible to consider by analysis alone the possible interactions between the architecture features, the application and the compiler. It is critical to employ a compiler-in-the-loop exploration, where the architectural changes are *made visible to and exploited by the compiler* to provide meaningful, quantitative feedback to the designer during architectural exploration.

By using a more systematic approach, where the designer can use the application information to customize the architecture, providing the architectural features to the compiler and rapidly evaluate different architectures early in the design process may significantly improve the design turn-around time. In this book we present an approach that simultaneously performs hardware customization of the memory architecture, together with software retargeting of the memory-aware compiler optimizations. This approach can significantly improve the memory system performance for varied power and cost profiles for programmable embedded systems.

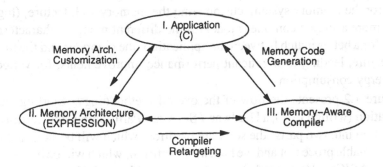

Figure 1.1. The interaction between the Memory Architecture, the Application and the Memory-Aware Compiler

Let us now examine our proposed memory system exploration approach. Figure 1.1 depicts three aspects of the memory sub-system that contribute towards the programmable embedded system's overall behavior: (I) the Application, (II) the Memory Architecture, and (III) the Memory Aware Compiler.

(I) The Application, written in C, contains a varied set of data structures and access patterns, characterized by different types of locality, storage and transfer requirements.

(II) One critical ingredient necessary for Design Space Exploration, is the ability to describe the memory architecture in a common description language. The designer or an exploration "space-walker" needs to be able to modify this description to reflect changes to the processor-memory architecture during Design Space Exploration. Moreover, this language needs to be understood by the different tools in the exploration flow, to allow interaction and inter-operability in the system. In our approach, the Memory Architecture, represented in an Architectural Description Language (such as EXPRESSION [HGG+99, MGDN01]), contains a description of the processor-memory architecture, including the memory modules (such as DRAMs, caches, stream buffers, DMAs, etc.), their connectivity and characteristics.

(III) The Memory-Aware Compiler uses the memory architecture description to efficiently exploit the features of the memory modules (such as access modes, timings, pipelining, parallelism). It is crucial to consider the interaction between all the components of the embedded system early during the design process. Designers have traditionally explored various characteristics of the processor, and optimizing compilers have been designed to exploit special architectural features of the CPU (e.g., detailed pipelining information). However, it is also important to explore the design space of Memory Architecture with memory-library-aware compilation tools that explicitly model and exploit the high-performance features of such diverse memory modules. Indeed, particularly for the memory system, customizing the memory architecture, (together with a more accurate compiler model for the different memory characteristics) allows for a better match between the application, the compiler and the memory architecture, leading to significant performance improvements, for varied cost and energy consumption.

Figure 1.2 presents the flow of the overall methodology. Starting from an application (written in C), a Hardware/ Software Partitioning step partitions the application into two parts: the software partition, which will be executed on the programmable processor and the hardware partition, which will be implemented through ASICs. Prior work has extensively addressed Hardware/ Software partitioning and co-design [GVNG94, Gup95]. This book concentrates mainly on the Software part of the system, but also discusses our approach in the context of a Hardware/Software architecture (Section 4.4).

The application represents the starting point for our memory exploration. After estimating the memory requirements, we use a memory/connectivity IP library to explore different memory and connectivity architectures (APEX

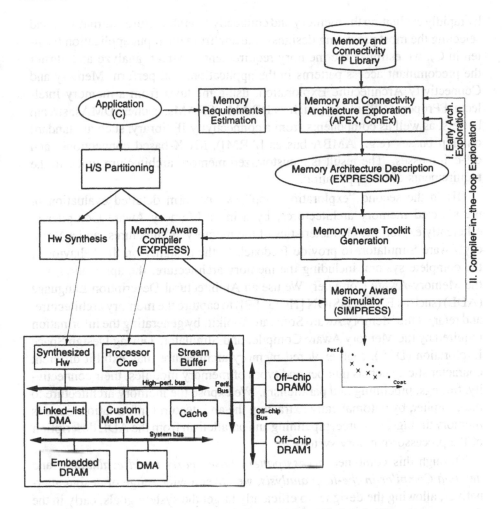

Figure 1.2. Our Hardware/Software Memory Exploration Flow

[GDN01b] and ConEx [GDN02]). The memory/connectivity architectures selected are then used to generate the compiler/simulator toolkit, and produce the pareto-like configurations in different design spaces, such as cost/performance and power. The resulting architecture in Figure 1.2 contains the programmable processor, the synthsized ASIC, and an example memory and connectivity architecture.

We explore the memory system designs following two major "exploration loops": (I) Early Memory Architecture Exploration, and (II) Compiler-in-the-loop Memory Exploration.

(I) In the first "exploration loop" we perform early Memory and Connectivity Architecture Exploration based on the access patterns of data in the application, by rapidly evaluating the memory and connectivity architecture alternatives, and selecting the most promising designs. Starting from the input application (written in C), we estimate the memory requirements, extract, analyze and cluster the predominant access patterns in the application, and perform Memory and Connectivity Architecture Exploration, using modules from a memory Intellectual Property (IP) library, such as DRAMs, SRAMs, caches, DMAs, stream buffers, as well as components from a connectivity IP library, such as standard on-chip busses (e.g., AMBA busses [ARM]), MUX-based connections, and off-chip busses. The result is a customized memory architecture tuned to the requirements of the application.

(II) In the second "exploration loop", we perform detailed evaluation of the selected memory architectures, by using a Memory Aware Compiler to efficiently exploit the characteristics of the memory architectures, and a Memory Aware Simulator, to provide feedback to the designer on the behavior of the complete system, including the memory architecture, the application, and the Memory-Aware Compiler. We use an Architectural Description Language (ADL) (such as EXPRESSION [HGG+99]) to capture the memory architecture, and retarget the Memory Aware Software Toolkit, by generating the information required by the Memory Aware Compiler and Simulator. During Design Space Exploration (DSE), each explored memory architecture may exhibit different characteristics, such as number and types of memory modules, their connectivity, timings, pipelining and parallelism. We expose the memory architecture to the compiler, by automatically extracting the architectural information, such as memory timings, resource, pipelining and parallelism from the ADL description of the processor-memory system.

Through this combined *access pattern based early rapid evaluation*, and *detailed Compiler-in-the-loop analysis*, we cover a wide range of design alternatives, allowing the designer to efficiently target the system goals, early in the design process without simulating the full design space.

Hardware/Software partitioning and codesign has been extensively used to improve the performance of important parts of the code, by implementing them with special purpose hardware, trading off cost of the system against better behavior of the computation [VGG94, Wol96a]. It is therefore important to apply this technique to memory accesses as well. Indeed, by moving the most active access patterns into specialized memory hardware (in effect creating a set of "memory coprocessors"), we can significantly improve the memory behavior, while trading off the cost of the system. We use a library of realistic memory modules, such as caches, SRAMs, stream buffers, and DMA-like memory modules that bring the data into small FIFOs, to target widely used data structures, such as linked lists, arrays, arrays of pointers, etc.

This two phase exploration methodology allows us to explore a space significantly larger than traditionally considered. Traditionally, designers have addressed the processor-memory gap by using simple cache hierarchies, and mainly relying on the designer's intuition in choosing the memory configuration. Instead our approach allows the designer to systematically explore the memory design space, by selecting memory modules and the connectivity configuration to match the access patterns exhibited by the application. The designer is thus able to select the most promising memory and connectivity architectures, using diverse memory modules such as DRAMs, SRAMs, caches, stream buffers, DMAs, etc. from a memory IP library, and standard connectivity components from a connectivity IP library.

1.3 Book Organization

The rest of this book is organized as follows:

Chapter 2: Related Work. We outline previous and related work in the domain of memory architecture exploration and optimizations for embedded systems.

Chapter 3: Early Memory Size Estimation. In order to drive design space exploration of the memory sub-system, we perform early estimation of the memory size requirements for the different data structures in the application.

Chapter 4: Memory Architecture Exploration. Starting from the most active access patterns in the embedded application, we explore the memory and connectivity architectures early during the design flow, evaluating and selecting the most promising design alternatives, which are likely to best match the cost, performance, and power goals of the system. These memory and connectivity components are selected from existing Intellectual Property (IP) libraries.

Chapter 5: Memory-aware Compilation. Contemporary memory components often employ special access modes (e.g., page-mode and burst-mode) and organizations (e.g., multiple banks and interleaving) to facilitate higher memory throughput. We present an approach that exposes such information to the Compiler through an Architecture Description Language (ADL). We describe how a memory-aware compiler can exploit the detailed timing and protocols of these memory modules to hide the latency of lengthy memory operations and boost the peformance of the applications.

Chapter 6: Experiments. We present a set of experiments demonstrating the utility of our Hardware/Software Memory Customization approach.

Chapter 7: Summary and Future Work. We present our conclusions, and possible future directions of research, arising from this work.

Chapter 2

RELATED WORK

In this chapter we outline previous approaches related to memory architecture design. Whereas there has been a very large body of work on the design of memory subsystems we focus our attention on work related to customization of the memory architecture for specific application domains. In relation to memory customization there has been work done in four main domains: (I) High-level synthesis, (II) Cache locality optimizations, (III) Computer Architecture, and (IV) Disk file systems and databases. We briefly describe these approaches; detailed comparisons with individual techniques presented in this book are described in ensuing chapters. In the context of embedded and special-purpose programmable architectures, we also outline work done in heterogeneous memory architectures and provide some examples.

2.1 High-Level Synthesis

The topic of memory issues in high-level synthesis has progressed from considerations in register allocation, through issues in the synthesis of foreground and background memories. In the domain of High-Level Synthesis, Catthoor et al. [CWG+98] address multiple problems in the memory design, including source level transformations to massage the input application, and improve the overall memory behavior, and memory allocation. De Greef et al. [DCD97] presents memory size minimization techniques through code reorganization and in-place mapping. They attempt to reuse the memory locations as much as possible by replacing data which is no longer needed with newly created values. Such memory reuse requires the alteration of the addressing calculation (e.g., in the case of arrays), which they realize through code transformations. Balasa et al. [BCM95] present memory estimation and allocation approaches for large multi-dimensional arrays for non-procedural descriptions. They determine the loop order and code schedule which results in the lowest memory requirement.

Wuytack et al. [WCdJ+96] present an approach to manage the memory bandwidth by increasing memory port utilization, through memory mapping and code reordering optimizations. They perform memory allocation by packing the data structures according to their size and bitwidth into memory modules from a library, to minimize the memory cost.

Bakshi et al. [BG95] perform memory exploration, combining memory modules using different connectivity and port configurations for pipelined DSP systems.

In the context of custom hardware synthesis, several approaches have been used to model and exploit memory access modes. Ly et. al. [LKMM95] use behavioral templates to model complex operations (such as memory reads and writes) in a CDFG, by enclosing multiple CDFG nodes and fixing their relative schedules (e.g., data is asserted one cycle after address for a memory write operation).

Panda et al. [PDN99] have addressed customization of the memory architecture targeting different cache configurations, or alternatively using on-chip scratch pad SRAMs to store data with poor cache behavior. Moreover, Panda et. al. [PDN98] outline a pre-synthesis approach to exploit efficient memory access modes, by massaging the input application (e.g., loop unrolling, code reordering) to better match the behavior to a DRAM memory architecture exhibiting page-mode accesses. Khare et. al. [KPDN98] extend this work to Synchronous and RAMBUS DRAMs, using burst-mode accesses, and exploiting memory bank interleaving.

Recent work on interface synthesis [COB95], [Gup95] present techniques to formally derive node clusters from interface timing diagrams. These techniques can be applied to provide an abstraction of the memory module timings required by the memory aware compilation approach presented in Chapter 5.

2.2   Cache Optimizations

Cache optimizations improving the cache hit ratio have been extensively addressed by both the embedded systems community ([CWG+98], [PDN99]) and the traditional architecture/compiler community ([Wol96a]). Loop transformations (e.g., loop interchange, blocking) have been used to improve both the temporal and spatial locality of the memory accesses. Similarly, memory allocation techniques (e.g., array padding, tiling) have been used in tandem with the loop transformations to provide further hit ratio improvement. However, often cache misses cannot be avoided due to large data sizes, or simply the presence of data in the main memory (compulsory misses). To efficiently use the available memory bandwidth and minimize the CPU stalls, it is crucial to aggressively schedule the loads associated with cache misses[1].

Pai and Adve [PA99] present a technique to move cache misses closer together, allowing an out-of-order superscalar processor to better overlap these misses (assuming the memory system tolerates a large number of outstanding misses).

The techniques we present in this book are complementary to the previous approaches: we overlap cache misses with cache hits to a different cache line. That is, while they cluster the cache misses to fit into the same superscalar instruction window, we perform static scheduling to hide the latencies.

Cache behavior analysis predicts the number/moment of cache hits and misses, to estimate the performance of processor-memory systems [AFMW96], to guide cache optimization decisions [Wol96a], to guide compiler directed prefetching [MLG92] or more recently, to drive dynamic memory sub-system reconfiguration in reconfigurable architectures [JCMH99], [VTG+99]. The approach presented in this book uses the cache locality analysis techniques presented in [MLG92], [Wol96a] to recognize and isolate the cache misses in the compiler, and then schedule them to better hide the latency of the misses.

2.3   Computer Architecture

In the domain of Computer Architecture, [Jou90], [PK94] propose the use of hardware stream buffers to enhance the memory system performance. Reconfigurable cache architectures have been proposed recently [VTG+99] to improve the cache behavior for general purpose processors, targeting a large set of applications.

In the embedded and general purpose processor domain, a new trend of instruction set modifications has emerged, targeting explicit control of the memory hierarchy through, for instance, prefetch, cache freeze, and evict-block operations (e.g., TriMedia 1100, StrongArm 1500, IDT R4650, Intel IA 64, Sun UltraSPARC III, etc. [hot]). For example, Harmsze et. al. [HTvM00] present an approach to allocate and lock the cache lines for stream based accesses, to reduce the interference between different streams and random CPU accesses, and improve the predictability of the run-time cache behavior.

Another approach to improve the memory system behavior used in general purpose processors is data prefetching. Software prefetching [CKP91], [GGV90], [MLG92], inserts prefetch instructions into the code, to bring data into the cache early, and improve the probability it will result in a hit. Hardware prefetching [Jou90], [PK94] uses hardware stream buffers to feed the cache with data from the main memory. On a cache miss, the prefetch buffers provide the

---

[1]In the remaining we refer to the scheduling of such loads as "scheduling of cache misses"

required cache line to the cache faster than the main memory, but comparatively slower than the cache hit access.

In the domain of programmable SOC architectural exploration, recently several efforts have used Architecture Description Languages (ADLs) to drive generation of the software toolchain (compilers, simulators, etc.) ([HD97], [Fre93], [Gyl94], [HGG+99], [LM98]). However, most of these approaches have focused primarily on the processor and employ a generic model of the memory subsystem. For instance, in the Trimaran compiler [Tri97], the scheduler uses operation timings specified on a per-operation basis in the MDes ADL to better schedule the applications. However they use fixed operation timings, and do not exploit efficient memory access modes. Our approach uses EXPRESSION [HGG+99], a memory-aware ADL that explicitly provides a detailed memory timing to the compiler and simulator.

The work we present in this book differs significantly from all the related work in that we simultaneously customize the memory architecture to match the access patterns in the application, while retargeting the compiler to exploit features of the memory architecture. Such an approach allows the system designer to explore a wide range of design alternatives, and significantly improve the system performance and power for varied cost configurations. Detailed comparison with individual approaches are presented in each chapter that follows in this book.

2.4   Disk File Systems

The topic of memory organization for efficient access of database objects has been studied extensively in the past. In the file systems domain, there have been several approaches to improve the file system behavior based on the file access patterns exhibited by the application. Patterson et. al. [PGG+95] advocate the use of hints describing the application access pattern to select particular prefetching and caching policies in the file system. Their work supports sequential accesses and an explicit list of accesses, choosing between prefetching hinted blocks, caching hinted blocks, and caching recently used un-hinted data. Their informed prefetching approach generates 20% to 83% performance improvement, while the informed caching generates a performance improvement of up to 42%. Parsons et. al [PUSS] present an approach allowing the application programmer to specify the file I/O parallel behavior using a set of templates which can be composed to form more complex access patterns. They support I/O templates such as meeting, log, report, newspaper, photocopy, and each may have a set of attributes (e.g., ordering attributes: ordered, relaxed and chaotic). The templates, described as an addition to the application source code, improve the performance of the parallel file system.

Kotz et. al [PEK+95b] present an approach to characterize the I/O access patterns for typical multiprocessor workloads. They classify them according to sequentiality, I/O request sizes, I/O skipped interval sizes, synchronization, sharing between processes, and time between re-writes and re-reads. Sequentiality classifies sequential and consecutive access patterns (sequential when the next access is to an offset larger than the current one, and consecutive when the next access is to the next offset), read/only, write/only and read/write. They obtain significant performance improvements, up to 16 times faster than traditional file systems, and using up to 93% of the peak disk bandwidth.

While the design of high performance parallel file systems depends on the understanding of the expected workload, there have been few usage studies of multiprocessor file system patterns. Purakayashta et. al [PEK+95a] characterize access patterns in a file system workload on a Connection Machine CM-5 to fill in this gap. They categorize the access patterns based on request size, sequentiality (sequential vs. consecutive), request intervals (number of bytes skipped between requests), synchronization (synchronous-sequential, asynchronous, local-independent, synchronous-broadcast, global-independent), sharing (concurrently shared, write shared), time between re-writes and re-reads, and give various recommendations for optimizing parallel file system design.

The idea at the center of these file system approaches using the access patterns to improve the file system behavior can be extrapolated to the memory domain. In our approach we use the memory accesses patterns to customize the memory system, and improve the match between the application and the memory architecture.

2.5   Heterogeneous Memory Architectures

Memory traffic patterns vary significantly between different applications. Therefore embedded system designers have long used heterogeneous memory architectures in order to improve the system behavior.

The recent trend towards low-power architectures further drive the need for exploiting customized memory subsystems that not only yield the desired performance, but also do so within an energy budget. Examples of such heterogeneous memory architectures are commonly found in special-purpose programmable processors, such as multimedia processors, network processors, DSP and even in general purpose processors; different memory structures, such as on-chip SRAMs, FIFOs, DMAs, stream-buffers are employed as an alternative to traditional caches and off-chip DRAMs. However, designers have relied mainly on intuition and previous experience, choosing the specific architecture in an ad-hoc manner. In the following, we present examples of memory architecture customization for the domain of network processors as well as other contemporary heterogeneous memory architectures.

2.5.1   Network Processors

Applications such as network processing place tremendous demands on throughput; typically even high-speed traditional processors cannot keep up with the high speed requirement. For instance, on a 10Gb/s (OC-192) link, new packets arrive every 35ns. Within this time, each packet has to be verified, classified, modified, before being delivered to the destination, requiring hundreds of RISC instructions, and hundreds of bytes of memory traffic per packet. We present examples of some contemporary network processors that employ heterogeneous memory architectures.

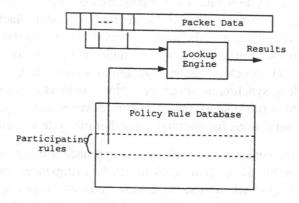

Figure 2.1.   Packet Classification in Network Processing

For instance, an important bottleneck in packet processing is packet classification [IDTS00]. As shown in Figure 2.1, packet classification involves several steps: first the different fields of the incoming packet are read; then, for each field a look-up engine is used to match the field to the corresponding policy rules extracted from the policy rule database, and finally the result is generated. Since the number of rules can be quite large (e.g., 1000 rules to 100,000 rules for large classifiers [BV01]), the number of memory accesses required for each packet is significant. Traditional memory approaches, using off-chip DRAMs, or simple cache hierarchies are clearly not sufficient to sustain this memory bandwidth requirement. In such cases, the use of special-purpose memory architectures, employing memory modules such as FIFOs, on-chip memories, and specialized transfer units are crucial in order to meet the deadlines.

Indeed, contemporary Network Processor implementations use different memory modules, such as on-chip SRAMs, FIFOs, CAMs, etc. to address the memory bottleneck. For instance the Intel Network Processor (NP) [mpr] contains Content Addressable Memories (CAMs) and local SRAM memories, as well as multiple register files (allowing communication with neighboring processors as well as with off-chip memories), to facilitate two forms of parallelism: (I) Functional pipelining, where packets remain resident in a single NP while several different functions are performed, and (II) Context pipelining, where packets move from one micro-engine to another with each micro-engine performing a single function on every packet in the stream. Similarly, the ClassiPI router engine from PMCSierra [IDTS00] employs a CAM to store a database of rules, and a FIFO to store the search results.

Another example of a network processor employing heterogeneous memory organizations is the Lexra NetVortex PowerPlant Network Processor [mpr] that uses a dual-ported SRAM together with a block-transfer engine for packet receive and transmit, and block move to/from shared memory. Due to the fact that networking code exhibits poor locality, they manage in software the on-chip memory instead of using caches.

2.5.2   Other Memory Architecture Examples

DSP, embedded and general purpose processors also use various memory configurations to improve the performance and power of the system. We present some examples of current architectures that employ such non-traditional memory organizations.

Digital Signal Processing (DSP) architectures have long used memory organizations customized for stream-based access. Newer, hybrid DSP architectures continue this trend with the use of both Instruction Set Architecture (ISA) level and memory architecture customization for DSP applications. For instance, Motorola's Altivec contains prefetch instructions, which command 4 stream channels to start prefetching data into the cache. The TI C6201 DSP Very Long Instruction Word (VLIW) processor contains a fast local memory, a off-chip main memory, and a Direct Memory Access (DMA) controller which allows transfer of data between the on-chip local SRAM and the off-chip DRAM. Furthermore page and burst mode accesses allow faster access to the DRAM.

In the domain of embedded and special-purpose architectures, memory subsystems are customized based on the characteristics and data types of the applications. For instance, Smart MIPS, the MIPS processor targeting Smart Cards, provides reconfigurable instruction and data scratch pad memory. Determining from an application what data structures are important to be stored on chip and configuring the processor accordingly is crucial for achieving the desired performance and power target. Aurora VLSI's DeCaff [mpr] Java accelerator uses a stack as well as a variables memory, backed by a data cache. The MAP-CA [mpr] from Equator multimedia processor uses a Data Streamer with an 8K SRAM buffer to transfer data between the CPU, coding engine, and video memory.

Heterogeneous memory organizations are also beginning to appear in mainstream general-purpose processors. For instance, SUN UltraSparc III [hot] uses prefetch caches, as well as on-chip memories that allow a software-controlled cache behavior, while PA 8500 from HP has prefetch capabilities, providing instructions to bring the data earlier into the cache, to insure a hit.

2.6   Summary

In this chapter we briefly surveyed several approaches that employ customized and heterogeneous memory architectures and organizations, in order to achieve the desired performance or power budgets of specific application domains.

The trend of such customization, although commonly used for embedded, or domain specific architectures, will continue to influence the design of newer programmable embedded systems. System architects will need to decide which groups of memory accesses deserve decoupling from the computations, and customization of both the memory organization itself, as well as the software interface (e.g., ISA-level) modifications to support efficient memory behavior. This leads to the challenging tasks of decoupling critical memory accesses from the computations, scheduling such accesses (in parallel or pipelined) with the computations, and the allocation of customized (special-purpose) memory transfer and storage units to support the desired memory behavior. Typically such tasks have been performed by system designers in an ad-hoc manner, using intuition, previous experience, and limited simulation/exploration. Consequently many feasible and interesting memory architectures are not considered. In the following chapters we present a systematic strategy for exploration of this memory architecture space, giving the system designer improved confidence in the choice of early, memory architectural decisions.

Chapter 3

EARLY MEMORY SIZE ESTIMATION [†]

3.1 Motivation

The high level design space exploration step is critical for obtaining a cost effective implementation. Decisions at this level have the highest impact on the final result.

A large class of applications (such as multimedia, DSP) exhibit complex array processing. For instance, in the algorithmic specifications of image and video applications the multidimensional variables (signals) are the main data structures. These large arrays of signals have to be stored in on-chip and off-chip memories. In such applications, memory often proves to be the most important hardware resource. Thus it is critical to develop techniques for early estimation of memory resources.

Early memory architecture exploration involves the steps of allocating different memory modules, partitioning the initial specification between different memory units, together with decisions regarding the parallelism provided by the memory system; each memory architecture thus evaluated exhibits a distinct cost, performance and power profile, allowing the system designer to trade-off the system performance against cost and energy consumption. To drive this process, it is important to be able to efficiently predict the memory requirements for the data structures and code segments in the application. Figure 3.1 shows the early memory exploration flow of our overall methodology outlined in Figure 1.2. As shown in Figure 3.1, during our exploration approach memory size estimates are required to drive the exploration of memory modules.

---

*Prof. Florin Balasa (University of Illinois, Chicago) contributed to the work presented in this chapter

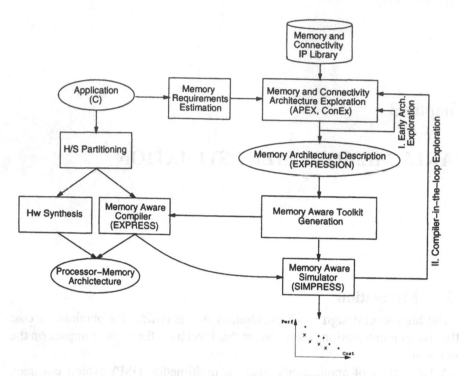

Figure 3.1.   Our Hardware/Software Memory Exploration Flow

We present a technique for memory size estimation, targeting procedural specifications with multidimensional arrays, containing both instruction level (fine-grain) and coarse- grain parallelism [BENP93], [Wol96b]. The impact of parallelism on memory size has not been previously studied in a consistent way. Together with tools for estimating the area of functional units and the performance of the design, our memory estimation approach can be used in a high level exploration methodology to trade-off performance against system cost.

This chapter is organized as follows. Section 3.2 defines the memory size estimation problem. Our approach is presented in Section 3.3. In Section 3.4 we discuss the influence of parallelism on memory size. Our experimental results are presented in Section 3.5. Section 3.6 briefly reviews some major results obtained in the field of memory estimation, followed by a summary in Section 3.7.

3.2   Memory Estimation Problem

We define the problem of memory size estimation as follows: given an input algorithmic specification containing multidimensional arrays, what is the number of memory locations necessary to satisfy the storage requirements of the system?

The ability to predict the memory characteristics of behavioral specifications without synthesizing them is vital to producing high quality designs with reasonable turnaround. During the HW/SW partitioning and design space exploration phase the memory size varies considerably. For example, in Figure 3.4, by assigning the second and third loop to different HW/SW partitions, the memory requirement changes by 50% (we assume that array a is no longer needed and can be overwritten). Here the production of the array b increases the memory by 50 elements, without consuming values. On the other hand, the loop producing array c consumes 100 values (2 per iteration). Thus, it is beneficial to produce the array c earlier, and reuse the memory space made available by array a. By producing b and c in parallel, the memory requirement is reduced to 100.

Our estimation approach considers such reuse of space, and gives a fast estimate of the memory size. To allow high- level design decisions, it is very important to provide good memory size estimates with reasonable computation effort, without having to perform complete memory assignment for each design alternative. During synthesis, when the memory assignment is done, it is necessary to make sure that different arrays (or parts of arrays) with non-overlapping lifetimes share the same space. The work in [DCD97] addresses this problem, obtaining results close to optimal. Of course, by increasing sharing between different arrays, the addressing becomes more complex, but in the case of large arrays, it is worth increasing the cost of the addressing unit in order to reduce the memory size.

Our memory size estimation approach uses elements of the polyhedral dataflow analysis model introduced in [BCM95], with the following major differences: (1) The input specifications may contain explicit constructs for parallel execution. This represents a significant extension required for design space exploration, and is not supported by any of the previous memory estimation /allocation approaches mentioned in Section 2. (2) The input specifications are interpreted procedurally, thus considering the operation ordering consistent with the source code. Most of the previous approaches operated on non-procedural specifications, but in practice a large segment of embedded applications market (e.g., GSM Vocoder, MPEG, as well as benchmark suites such as DSPStone [ZMSM94], EEMBC [Emb]) operate on procedural descriptions, so we consider it is necessary to accommodate also these methodologies.

Our memory size estimation approach handles specifications containing nested loops having affine boundaries (the loop boundaries can be constants or linear functions of outer loop indexes). The memory references can be multidimensional signals with (complex) affine indices. The parallelism is explicitly described by means of cobegin-coend and forall constructs. This parallelism could be described explicitly by the user in the input specification, or could be generated through parallelizing transformations on procedural code. We assume the input has the single-assignment property [CF91] (this could be generated through a preprocessing step).

The output of our memory estimation approach is a range of the memory size, defined by a lower- and upper-bound. The predicted memory size for the input application lies within this range, and in most of the cases it is close to the lower-bound (see the experiments). Thus, we see the lower bound as a prediction of the expected memory size, while the upper bound gives an idea of the accuracy of the prediction (i.e., the error margin). When the two bounds are equal, an "exact" memory size evaluation is achieved (by exact we mean the best that can be achieved with the information available at this step, without doing the actual memory assignment). In order to handle complex specifications, we provide a mechanism to trade-off the accuracy of predicting the storage range against the computational effort.

3.3   Memory Size Estimation Algorithm

Our memory estimation approach (called MemoRex) has two parts. Starting from a high level description which may contain also parallel constructs, the Memory Behavior Analysis (MBA) phase analyzes the memory size variation, by approximating the memory trace, as shown in Figure 3.2, using a covering bounding area. Then, the Memory Size Prediction (MSP) computes the memory size range, which is the output of the estimator. The backward dotted arrow in Figure 3.2 shows that the accuracy can be increased by subsequent passes.

The memory trace represents the size of the occupied storage in each logical time step during the execution of the input application. The continuous line in the graphic from Figure 3.2 represents such a memory trace. When dealing with complex specifications, we do not determine the exact memory trace due to the high computational effort required. A bounding area encompassing the memory trace - the shaded rectangles from the graphic in Figure 3.2 - is determined instead.

The storage requirement of an input specification is obviously the peak of the (continuous) trace. When the memory trace cannot be determined exactly, the approximating bounding area can provide the lower- and upper- bounds of the trace peak. This range of the memory requirement represents the result of our estimation approach.

The MemoRex algorithm (Figure 3.3) has five steps. Employing the terminology introduced in [vSFCM93], the first step computes the number of array elements produced by each definition domain and consumed by each operand domain. The definition/operand domains are the array references in the left/right hand side of the assignments. A definition produces the array elements (the array elements are created), while the last read to an array ele-

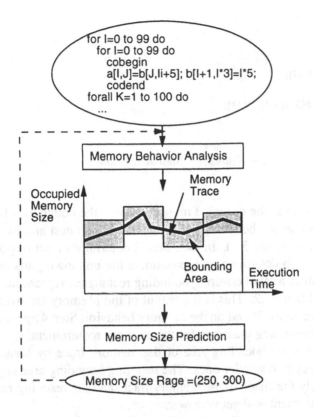

Figure 3.2. The flow of MemoRex approach

Memory Behavior Analysis
1. Perform data-dependence analysis
2. Compute memory size between loop nests
3. Determine bounding area for the memory trace

Memory Size Prediction
4. Determine memroy size range

Accuracy Increase
5. If more accuracy needed, split critical bounding areas, goto step 1

Figure 3.3. Outline of the MemoRex algorithm.

ment consumes the array element (the element is no longer needed, and can be potentially discarded).

```
for I=1 to 10 do
  a[I] = I;
for I=1 to 10 do
  b[I] = a[I] + a[11-I];
for I=1 to 5 do
  c[I] = a[2*I] + b[2*I] + b[2*I+1];
```

Figure 3.4.   Illustrative Example

Step 2 determines the occupied memory size at the boundaries between the loop nests (such as the boundary between the for loop nest and the forall loop in the code from Figure 3.2). In fact, Step 2 determines a set of points on the memory trace. To determine or approximate the unknown parts of the trace, Step 3 determines a set of covering bounding rectangles, represented as shaded rectangles in Figure 3.2. This is the output of the Memory Behavior Analysis part of our algorithm. Based on the memory behavior, Step 4 approximates the trace peak, determining the range for the memory requirement.

Step 5 refines the bounding area of the memory trace by breaking up the larger rectangles into smaller ones. The resulting bounding area approximates more accurately the shape of the memory trace, and the resulting range for the memory requirement will get narrower.

Each step of the MemoRex algorithm in Figure 3.3 is presented in the following; we employ for illustration the simple code in Figure 3.4 to present in detail the algorithm.

3.3.1   Data-dependence analysis

By studying the dependence relations between the array references in the code, this step determines the number of array elements produced (created) or consumed (killed) during each assignment.

The number of array elements produced by an assignment is given by the size of the corresponding definition domains. In the illustrative example, the number of array elements produced in the three loops are $Card\{a[I], 1 <= I <= 10\} = 10$, $Card\{b[I], 1 <= I <= 10\} = 10$, and $Card\{c[I], 1 <= I <= 5\} = 5$, respectively. In general though, the size of signal domains is more difficult to compute, for instance, when handling array references within the scope of loop nests and conditions: our approach employs the algorithm which determines the size of linearly bounded lattices, described in [BCM95].

On the other hand, the number of array elements consumed by an operand domain, is not always equal to the size of the operand domain, as some of the array elements may belong also to other operands from subsequent assignments.

For instance, only two array elements are consumed by the operand domain a[I] (i.e., a[7], a[9]) and three other array elements (a[1], a[3], a[5]) by the operand domain a[11-I], as the other a array elements of even index are read also by the operand a[2*I] in the third loop.

In general, the computation of "killed" signals is more complicated when dealing with loop nests and conditions. We perform the computation employing a symbolic time function for each assignment, which characterizes (in a closed form formula) when each array element is read, thus allowing us to find the last read of each array element, i.e., the domain consuming that element.

3.3.2 Computing the memory size between loop nests

For each nest of loops, the total number of memory locations produced/consumed is the sum of the locations produced/ consumed in each domain within that loop nest. The memory size after executing a loop nest is the sum of the memory size at the beginning of the loop nest, and the number of array elements produced minus the number of array elements consumed within the loop nest:

$$mem(end\_loop) = mem(begin\_loop) + sum(prod's) - sum(consmp's)$$

As shown in Figure 3.5 a, the memory size for our example is 0 at the beginning of the first loop, 10 after the execution of the first loop (because this loop produces 10 array elements and does not consume any), and 15 at the end of the second loop, as 10 new array elements are produced (b[1..10]), while the five odd-index array elements of a are no longer necessary.

3.3.3 Determining the bounding rectangles

Based on the information already acquired in Steps 1 and 2, our algorithm constructs bounding rectangles for each loop nest in the specification. These rectangles are built such that they cover the memory trace (see Figure 3.5 b). Thus, they characterize the behavior of the memory size for the portion of code under analysis.

We illustrate the construction of the bounding rectangles for the second loop in Figure 3.4. It is known from Step 1 that 10 array elements (b[1..10]) are produced, while the operand domain a[I] consumes 2 array elements, and the operand domain a[11-I] consumes 3. Since at most 7 out of the 10 assignments may not consume any values (the other 3 will consume at least 1 value), the maximum storage variation occurs, if the first 7 assignments generate one new value each, without consuming any, and all the consumptions occur later. Knowing from Step 2 that the memory size is 10 at the beginning of loop 2, it follows that the upper-bound of the memory size for this loop is 10+7=17 locations. With similar reasoning, one can conclude that during the execution of this loop, the memory trace could not go below 8 locations (see Figure 3.5

c). Thus, the bounding rectangle for this loop has the upper edge 17, and lower edge 8.

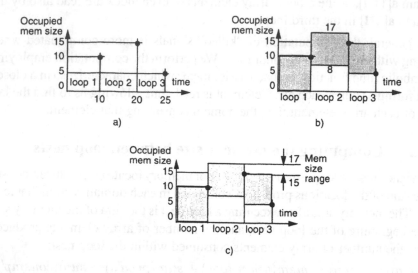

Figure 3.5.   Memory estimation for the illustrative example

3.3.4   Determining the memory size range

The memory requirement for a specification is the peak of the memory trace. Since the peak of the trace is contained within the bounding rectangles (along with the whole memory trace), the highest point among all the bounding rectangles represents an upper-bound of the memory requirement. For our illustrative example, the memory requirement will not exceed 17 - the highest upper-edge of the bounding rectangles (see Figure 3.5 c).

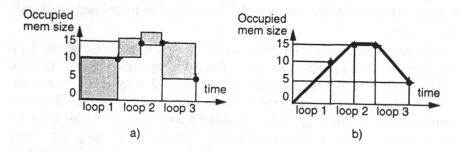

Figure 3.6.   (a) Accuracy refinement, (b) Complete memory trace

Since the memory size at the boundaries between the loop nests is known, the memory requirement will be at least the maximum of these values. The maximum memory size at the boundary points thus represents the lower-bound of the memory requirement. For our illustrative example, the memory requirement is higher than 15 (the lower dotted line in Figure 3.5c), which is the maximum of the values at the boundaries of the three loops (Figure 3.5a). Therefore, the actual memory requirement will be in the range [15..17]. This memory requirement range represents the result of the first pass of the algorithm. The last step of our algorithm decides whether a more accurate approximation is necessary, in such case initiating an additional pass.

3.3.5 Improving the estimation accuracy

If the current estimation accuracy is not satisfactory, for each loop nest whose rectangle may contain the memory trace peak (the upper-edge higher than the previous memory lower-bound), a split of the iteration space is performed (by gradually splitting the range of the iterators, thus fissioning the loop nest). The critical rectangles corresponding to these loop nests will be replaced in a subsequent pass of the estimation algorithm by two new rectangles covering a smaller area (Figure 3.6 a) and, therefore, following more accurately the actual memory trace, thus yielding a more accurate memory behavior analysis, and a more exact estimation. This process is iteratively repeated until a convenient estimation accuracy is reached. Figure 3.6 a shows the refined bounding area, and Figure 3.6 b presents the actual memory trace for our example.

3.4 Discussion on Parallelism vs. Memory Size

In the following we present a more complex example (Figure 3.7), which shows the utility of the estimation tool. This code contains a forall loop and multiple instructions executed in parallel (using cobegin-coend), assuming unlimited processing resources. We consider the behavior of this code in two cases: (I) Assuming that all "forall" loops are executed sequentially, and (II) Assuming a parallel execution model.

(I) Assuming the forall loop executes sequentially, we obtain the behavior in Figure 3.8, and after the second pass we determine that the memory size is between [396..398].

(II) By allowing the forall loop to be executed in parallel, the memory behavior becomes the one depicted in Figure 3.9, and the memory size is 300. Thus, the memory size for the parallel version of the code is 25% less than for the sequential case.

Having the parallel version require 25% less memory than the sequential one is a surprising result, since common sense would suggest that more parallelism in the code would need more memory. We have done some preliminary experiments in this direction, and they all seem to imply that more parallelism does not necessarily mean more memory. Moreover, we could not find (or produce) any

```
for I=0 to 97 do
  a[2*I,0] = 0; a[2*I+1,0] = 1;
  for J=0 to 99 do cobegin
    a[2*I,J+1] = a[2*I+1,J];
    a[2*I+1,J+1] = a[2*I,J];
  coend;
forall I=98 to 99 do
  a[2*I,0] = 0; a[2*I+1,0] = 1;
  for J=0 to 99 do cobegin
    a[2*I,J+1] = a[2*I+1,J];
    if(I != 99) a[2*I+1,J+1] = a[2*I,J];
    else a[2*I+1,J+1] = a[2*I,J] + a[2*I-2,J];
  coend;
for J=0 to 99 do
  for I=0 to 99 do cobegin
    a[2*I,J+101] = a[2*I+1,J+100];
    a[2*I+1,J+101] = a[2*I,J+100] + a[2*I,J+99];
  coend;
```

Figure 3.7. Input Specification with parallel instructions

Memory size = 398

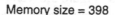

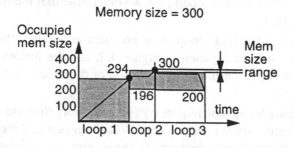

Figure 3.8. Memory behavior for example with sequential loop

Memory size = 300

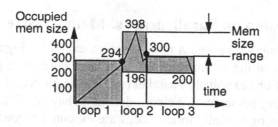

Figure 3.9. Memory behavior for example with *forall* loop

example where the most parallel version of the code needs more memory than the sequential versions. For most of the examples, the most parallel version had the same memory requirement as the "best" of the sequential ones. A possible explanation could be the fact that when instructions are executed in parallel, values are produced early, *but also consumed early*, and early consumption leads to low memory requirement.

3.5 Experiments

To estimate the effectiveness of our approach, we compared our algorithm against a memory estimation tool based on symbolic execution, which assigns the signals to memory on a scalar basis to maximize sharing [GBD98]. We ran both algorithms on a SPARCstation 5, on 7 application kernels: image compression algorithm (Compress), linear recurrence solver (Linear), image edge enhancement (Laplace), successive over-relaxation algorithm [PTVF92] (SOR), two filters (Wavelett, Low-pass), and red-black Gauss-Seidel method [PTVF92]. These examples are typical in image and video processing. Some of them (e.g., SOR, Wavelett, G-S) exhibit complex affine indices and conditionals.

The results of our tests are displayed in Table 3.1. Columns 2 and 3 show the memory requirement and the computation time obtained with the symbolic execution method. This memory size represents the optimal in terms of locations sharing. Column 4 represents the memory size estimate obtained with our algorithm (note that this is a lower-bound for the optimal value). For all the examples, the memory prediction is very close to the optimal value from column 2. The upper bound from Column 5 is used mainly as a measure of the confidence in the prediction. If this value is closer to the lower-bound estimate, the prediction is more accurate. Column 6 represents the percentile range.

$$100 * (upper\_bound - lower\_bound)/upper\_bound.$$

The lower this value, the more confidence we can have in the estimation. For most applications, high confidence is obtained within very reasonable time. A 0 percentile range means that the lower-bound is equal to the upper-bound, producing the exact value. When there is some slack between the two, the optimal value is usually very close to the lower-bound (the memory estimate), but in worst case, depending on the memory behavior complexity, it can be anywhere within that range.

For the Wavelett example, even though we obtained a very good estimate (48002 vs. 48003) from the first pass, we needed 3 passes to reduce the percentile range and increase the confidence in the estimation, due to the complexity of the memory behavior (the memory trace inside the loops is very irregular).

Application	Symbolic exec.		MemoRex algorithm			
	Exact mem size	Time [s]	Est mem size	Upper-bound	%	Time [s]
Compress	10000	133	10000	10000	0	0.01
Linear	20002	389	20002	20002	0	0.02
Laplace	10404	87	10404	10404	0	1
SOR	38265	205	38265	38265	0	0.01
Wavelett	48003	1265	48002	72002	33	0.02
			48002	54002	11	0.15
			48002	51002	5	0.42
Low-pass	10100	384	10099	10297	1	0.27
Gauss-Seidel	13870	197	13870	13917	0	5

Table 3.1. Experimental Results.

3.6   Related Work

One of the earliest approaches to estimation of scalar memory elements is the left edge algorithm [KP87], which assigns the scalar variables to registers. This approach is not suited for multidimensional signal processing applications, due to the prohibitive computational effort.

One of the earliest approaches of handling arrays of signals is based on clustering the arrays into memory modules such that a cost function is minimized [RGC94]. The possibility of signals with disjoint life times to share common storage locations is however ignored, the resulting memory requirements often significantly exceeding the actual storage needed. More recently, [ST97] proposed a more refined array clustering, along with a technique for binding groups of arrays to memory modules drawn from a given library. However, the technique does not perform in-place mapping within an array.

Approaches which deal with large multidimensional arrays operate on non-procedural [BCM95] and stream models [LMVvdW93]. Non-procedural specifications do not have enough information to estimate accurately the memory size, since by changing the instruction sequence, large memory variations are produced. For example, assuming the code in Figure 3.10 (a) is non-procedural, the memory size could vary between 100 and 150 locations, as in Figure 3.10 (b). [VSR94] uses a data-flow oriented view, as [BCM95], but has good results for simpler specifications (constant loop bounds, simpler indices). [vSFCM93] modified the loop hierarchy and the execution sequence of the source code, by placing polyhedrons of signals derived from the operands in a common space and determining an ordering vector in that space. None of the above techniques addresses specifications containing explicit parallelism.

Zhao et al. [ZM99] present an approach for memory size estimation for array computations based on live variable analysis, and integer point counting for parametrized polytopes. They show that it is enough to compute the number of live variables for only one instruction in each innermost loop to determine the minimum memory requirement. However, the live variable analysis is performed for each iteration of the loops, making it computationally intensive for large, multi-dimensional arrays.

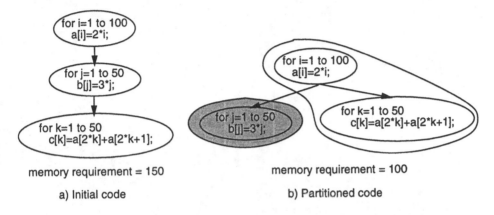

memory requirement = 150   memory requirement = 100

a) Initial code   b) Partitioned code

Figure 3.10.   Memory Size Variation during partitioning/parallelization

Other memory management approaches use the access frequencies to balance the port usage [PS97], and to optimize the partitioning between on-chip scratch-pad and cache-accessed memories [PDN97]. However, they do not consider the memory size.

3.7   Summary

In this chapter, we presented a technique for estimating the memory size for multidimensional signal processing applications, as part of our design space exploration environment. Different from previous approaches, we have addressed this problem considering that the algorithmic specifications written in a procedural style may also contain explicit parallel constructs. Even if the initial input does not contain explicit parallelism, partitioning and design space exploration may introduce explicit parallelism in an attempt to achieve higher performance.

Our experiments on typical video and image processing kernels show close to optimal results with very reasonable time. The experimental results obtained by our approach have been compared to a brute-force exact computation of the memory size, implemented using symbolic simulation.

for partially fixed polytopes. They allow for its enough to compute the number of live variables for only one iteration in each innermost loop to determine the maximum memory requirement. However, the extra variable analysis is performed for each iteration of the loop, making it computationally intensive for large, multi-dimensional arrays.

Figure 3.16: Memory size estimation during partitioning optimization.

Other memory management approaches use the measurement to estimate the partitioning [PS02], and to optimize the partitioning between on-chip scratch-pad and cache memories (PDN97). However, they do not consider the memory size.

3.7   Summary

In this report, we present a technique for estimating the memory size for multi-dimensional signal processing applications, in particular our design space exploration environment. Unlike our team previous approaches, we have all data as this problem considering that the algorithmic specifications within an architectural style may also contain explicit parallel operations. Even at the highest input. These notions exploit parallelism, partitioning and design space exploration by hence explicit parallelism in an attempt to achieve higher performance.

Our experiments on typical video applications, image processing kernels show close to optimal results with very reasonable time. The estimation results obtained by our approach have been compared to a more exact configuration of the memory size, implemented in a symbolic simulator.

Chapter 4

EARLY MEMORY AND CONNECTIVITY ARCHITECTURE EXPLORATION

4.1 Motivation

Traditionally, designers have attempted to improve memory behavior by exploring different cache configurations, with limited use of more special purpose memory modules such as stream buffers [Jou90]. However, while real-life applications contain a large number of memory references to a diverse set of data structures, a significant percentage of all memory accesses in the application are generated from a few instructions in the code. For instance, in Vocoder, a GSM voice coding application with 15K lines of code, 62% of all memory accesses are generated by only 15 instructions. Furthermore, these instructions often exhibit well-known, predictable access patterns. This presents a tremendous opportunity to customize the memory architecture to match the needs of the predominant access patterns in the application, and significantly improve the memory system behavior. Moreover, the cost, bandwidth and power footprint of the memory system is influenced not only by the memory modules employed, but also by the connectivity used to transfer the data between the memory modules and the CPU. While the configuration and characteristics of the memory modules are important, often the connectivity structure has a comparably large impact on the system performance, cost and power; thus it is critical to consider it early in the design flow. In this chapter we present an approach that couples the memory architecture exploration step (which extracts, analyzes, and clusters the most active memory access patterns in the application), with the connectivity exploration step (which evaluates a wide range of connectivity configurations using components from a connectivity IP library, such as standard on-chip busses, MUX-based connections, and off-chip busses). This coupled approach improves the performance of the system, for varying cost, and power consumption, allowing the designer to best tradeoff the different goals of the system. In Section 4.2 we present our Access Pattern based Memory Architecture Exploration approach, and in Section 4.3 we present our Connectivity Exploration approach.

4.2   Access Pattern Based Memory Architecture Exploration

Memory access patterns have long been analyzed in order to improve system performance. For instance, in the context of disk file systems [PUSS, PGG+95] there have been many approaches to employ the file access pattern to customize the file system to best match the access characteristics of the application. Likewise, many approaches have proposed customizing the processor through special purpose instructions and special functional units to target the predominant computations in embedded applications (such as MAC, FFT, etc.). However, none of the previous approaches has attempted to analyze the memory access patterns information in the application, with the goal of aggressively customizing the memory architecture. We use the access patterns to customize the memory architecture, employing modules from a memory IP library, to explore a wide range of cost, performance, and power designs. We use a heuristic to prune the design space of such memory customizations, and guide the search towards the designs with best cost/gain ratios, exploring a space well beyond the one traditionally considered.

In Section 4.2.1 we present the flow of our approach. In Section 4.2.2 we use a large real-life example to illustrate our approach and in Section 4.2.3 we present an outline of our Access Pattern based Memory Exploration (APEX) approach. In Section 4.2.4 we present a set of experiments that demonstrate the customization of the memory architecture for a set of large multimedia and scientific applications, and present exploration results showing the wide range of performance, power and cost tradeoffs obtained. In Section 4.2.5 we present the related work in the area of memory subsystem optimizations.

4.2.1   Our approach

Figure 4.1 presents the flow of our Access Pattern based Memory Exploration (APEX) approach. We start by extracting the most active access patterns from the input C application; we then analyze and cluster these access patterns according to similarities and interference, and customize the memory architecture by allocating a set of memory modules from a Memory Modules IP Library. We explore the space of these memory customizations by using a heuristic to guide the search towards the most promising cost/performance memory architecture tradeoffs. We prune the design space by using a fast time-sampling simulation to rule-out the non-interesting parts of the design space, and then fully simulate and determine accurate performance/power measures only for the selected memory architectures. After narrowing down the search to the most promising cost/performance designs, we allow the designer to best match the performance/power requirements of the system, by providing full cost/performance/power characteristics for the selected designs.

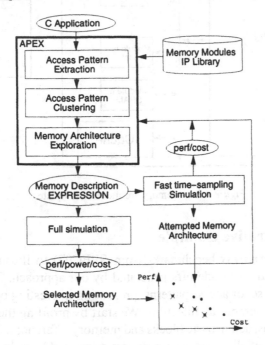

Figure 4.1. The flow of our Access Pattern based Memory Exploration Approach (APEX).

The basic idea is to target specifically the needs of the most active memory access patterns in the application, and customize a memory architecture, exploring a wide range of designs, that exhibit varied cost, performance, and power characteristics.

Figure 4.2 presents the memory architecture template. The memory access requests from the processor are routed to one of the memory modules 0 through n or to the cache, based on the address. The custom memory modules can read the data directly from the DRAM, or alternatively can go through the cache which is already present in the architecture, allowing access patterns which exhibit locality to make use of the locality properties of the cache. The custom memory modules implement different types of access patterns, such as stream accesses, linked-list accesses, or a simple SRAM to store hard-to-predict or random accesses. We use custom memory modules to target the most active access patterns in the application, while the remaining, less frequent access patterns are serviced by the on-chip cache.

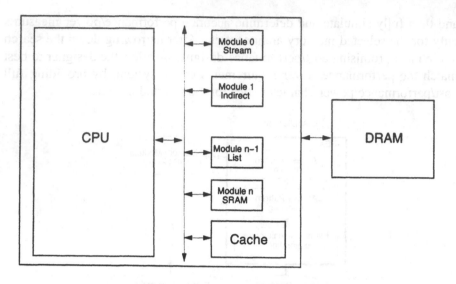

Figure 4.2. Memory architecture template.

4.2.2 Illustrative example

We use the *compress* benchmark (from SPEC95) to illustrate the performance, power and cost trade-offs generated by our approach. The benchmark contains a varied set of access patterns, presenting interesting opportunities for customizing the memory architecture. We start by profiling the application, to determine the most active basic blocks and memory references. In the *compress* benchmark, 40% of all memory accesses are generated by only 19 instructions.

By traversing the most active basic blocks, we extract the most active access patterns from the application. Figure 4.3 shows an excerpt of code from *compress*, containing references to 3 arrays: *htab*, *codetab*, and rmask. *htab* is a hashing table represented as an array of 69001 unsigned longs (we assume that both longs and ints are stored using 32 bits), *codetab* is an array of 69001 shorts, and rmask is an array of 9 characters. The sequence of accesses to *htab*, *codetab*, and rmask represent access patterns *ap1*, *ap2* and *ap3* respectively. The hashing table *htab* is traversed using the array *codetab* as an indirect index, and the sequence of accesses to the array *codetab* is generated by a self-indirection, by using the values read from the array itself as the next index. The sequence in which the array rmask is traversed is difficult to predict, due to a complex index expression computed across multiple functions. Therefore we consider the order of accesses as unknown. However, rmask represents a small table of coefficients, accessed very often.

Compress contains many other memory references exhibiting different types of access patters such as streams with positive or negative stride. We extract the most active access patterns in the application, and cluster them according

```
while ( ... )
...
... = htab[code];
code = codetab[code];
...
while ( ... )
...
... = rmask[r_off]
...

Access patterns:
    ap1 = htab[ap2]
    ap2 = codetabl[ap2]
    ap3 = rmask[unknown]
```

Figure 4.3. Example access patterns.

to similarity and interference. Since all the access patterns in a cluster will be treated together, we group together the access patterns which are compatible (for instance access patterns which are similar and do not interfere) in the hope that all the access patterns in a cluster can be mapped to one custom memory module.

Next, for each such access pattern cluster we allocate a custom memory module from the memory modules library. We use a library of parameterizable memory modules containing both generic structures such as caches and on-chip SRAMs, as well as a set of parameterizable custom memory modules developed for specific types of access patterns such as streams with positive, negative, or non-unit strides, indirect accesses, self-indirect accesses, linked-list accesses. The custom memory modules are based on approaches proposed in the general purpose computing domain [cC94, Jou90, RMS98], with the modification that the dynamic prediction mechanisms are replaced with the static compile-time analysis of the access patterns, and the prefetched data is stored in special purpose FIFOs.

We briefly describe one such example for illustration: for the sample access pattern *ap2* from *compress*, we use a custom memory module implementing self-indirect access pattern, while for the access pattern *ap3*, due to the small size of the array rmask, we use a small on-chip SRAM [PDN99]. Figure 4.4 presents an outline of the self-indirect custom memory module architecture used for the access pattern *ap2*, where the value read from the array is used as the index for the next access to the array. The base register stores the base address of the array, the index register stores the previous value which will be used as an index in the next access, and the small FIFO stores the stream of values read from the next memory level, along with the address tag used for write coherency. When the CPU sends a read request, the data is provided from the FIFO. The empty spot in the FIFO initiates a fetch from the next level memory to bring in the next data element. The adder computes the address for the next data element based on the base address and the previous data value. We assume that the base register is initialized to the base of the *codetab* array and the index register to the initial index through a memory mapped control register model (a store to the address corresponding to the base register writes the base address value into the register).

The custom memory modules from the library can be combined together, based on the relationships between the access patterns. For instance, the access pattern *ap1* uses the access pattern *ap2* as an index for the references. In such a case we use the self-indirection memory module implementing *ap2* in conjunction with a simple indirection memory module, which computes the sequence of addresses by adding the base address of the array *htab* with the values produced by *ap2*, and generate ap1=htab[ap2].

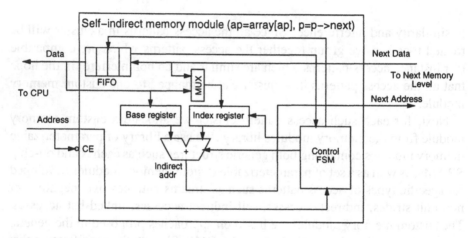

Figure 4.4.   Self-indirect custom memory module.

After selecting a set of custom memory modules from the library, we map the access pattern clusters to memory modules. Starting from the traditional memory architecture, containing a small cache, we incrementally customize access pattern clusters, to significantly improve the memory behavior. Many such memory module allocations and mappings are possible. Exploring the full space of such designs would be prohibitively expensive. In order to provide the designer with a spectrum of such design points without the time penalty of investigating the full space, we use a heuristic to select the most promising memory architectures, providing the best cost/performance/power tradeoffs.

For the *compress* benchmark we explore the design space choosing a set of 5 memory architectures which provide advantageous cost/performance tradeoffs. The overall miss rate of the memory system is reduced by 39%, generating a significant performance improvement for varied cost and power characteristics (we present the details of the exploration in Chapter 6). In this manner we can customize the memory architecture by extracting and analyzing the access patterns in the application, thus substantially improving the memory system behavior, and allowing the designer to trade off the different goals of the system.

4.2.3 The Access Pattern based Memory Exploration (APEX) Approach

Our Access Pattern based Memory Exploration (APEX) approach is a heuristic method to extract, analyze, and cluster the most active access patterns in the application, and customize the memory architecture, explore the design space to tradeoff the different goals of the system. It contains two phases: (I) Access pattern clustering and (II) Exploration of custom memory configurations.

4.2.3.1 Access Pattern Clustering

In the first phase of our approach, we extract the access patterns from the application, analyze and group them into access pattern clusters, according to their relationships, similarities and interferences. Figure 4.5 presents an outline of the access pattern extraction and clustering algorithm. The access pattern clustering algorithm contains 4 steps.

(1) We extract the most active access patterns from the input application. We consider three types of access patterns: (a) Access patterns which can be determined automatically by analyzing the application code, (b) Access patterns about which the user has prior knowledge, and (c) Access patterns that are difficult to determine, or are input-dependent.

(a) Often access patterns can be determined at compile time, using traditional compiler analysis. Especially in DSP and embedded systems, the access patterns tend to be more regular, and predictable at compile time (e.g., in video, image and voice compression).

First, we use profiling to determine the most active basic blocks in the application. For each memory reference in these basic blocks we traverse the use-def chains to construct the address expression, until we reach statically known variables, constants, loop indexes, or other access patterns. This closed form formula represents the access pattern of the memory reference. If all the elements in this expression are statically predictable, and the loop indexes have known bounds, the access pattern represented by this formula is predictable.

(b) In the case of well-known data structures (e.g., hashing tables, linked lists, etc.), or well-understood high-level concepts (such as the traversal algorithms in well-known DSP functions), the programmer has prior knowledge on the data structures and the access patterns. By providing this information in the form of assertions, he can give hints on the predominant accesses in the application. Especially when the memory references depend on variables

```
Procedure GenerateAccessPatternClusters
Input: Application in C and Access Pattern Assertions
Output: Access Pattern Clusters
begin
    1. Extract Access Patterns from application
    2. Build Access Pattern Graph APG(AP,Arcs)
    3. Build Access Pattern Compatibility Graph
            APCG(AP,CompatibilityArcs)
    4. Choose Cliques Of Compatibility Arcs to form
            Access Pattern Clusters
end
```

Figure 4.5.   Access Pattern Clustering algorithm.

which traverse multiple functions, indirections, and aliasing, and determining the access pattern automatically is difficult, allowing the user to input such readily available information, significantly improves the memory architecture customization opportunities.

(c) In the case of memory references that are complex and difficult to predict, or depend on input data, we treat them as random access patterns. While for such references it is often impossible to fully understand the access pattern, it may be useful to use generic memory modules such as caches or on-chip scratch pad memories, to exploit the locality trends exhibited.

(2) In the second step of the Access Pattern Clustering algorithm we build the Access Pattern Graph (APG), containing as vertices the most active access patterns from the application. The arcs in the APG represent properties such as similarity, interference, whether two access patterns refer to the same data structure, or whether an access pattern uses another access pattern as an index for indirect addressing, or pointer computation.

(3) Based on the APG, we build the Access Pattern Compatibility Graph (APCG), which has the same vertices as the APG (the access patterns), but the arcs represent compatibility between access patterns. We say two access patterns are *compatible*, if they can belong to the same access pattern cluster. For instance, access patterns that are similar (e.g., both have stream-like behavior), but which have little interference (are accessed in different loops) may share the same custom memory module, and it makes sense to place them in the same cluster. The meaning of the access pattern clusters is that all the access patterns in a cluster will be allocated to one memory module.

(4) In the last step of the Access Pattern Clustering algorithm, we find the cliques of fully connected subgraphs in the APCG compatibility graph. Each such clique represents an access pattern cluster, where all the access patterns

are compatible, according to the compatibility criteria determined from the previous step (a complete description of the compatibility criteria is presented in [GDN01c]). Each such access pattern cluster will be mapped in the following phase to a memory module from the library.

4.2.3.2   Exploring Custom Memory Configurations

In the second phase of the APEX approach, we explore the custom memory module implementations and access pattern cluster mappings, using a heuristic to find the most promising design points.

Figure 4.13 presents an outline of our exploration heuristic. We first initialize the memory architecture to contain a small traditional cache, representing the starting point of our exploration.

For each design point, the number of alternative customizations available is large, and fully exploring them is prohibitively expensive. For instance, each access pattern cluster can be mapped to custom memory modules from the library, or to the traditional cache, each such configuration generating a different cost/ performance/power tradeoff. In order to prune the design space, at each exploration step we first estimate the incremental cost and gain obtained by the further possible customization alternatives, then choose the alternative leading to the best cost/gain ratio for further exploration. Once a customization alternative has been chosen, we consider it the current architecture, and perform full simulation for the new design point. We then continue the exploration, by evaluating the further possible customization opportunities, starting from this new design point.

We tuned our exploration heuristic to prune out the design points with poor cost/ performance characteristics, guiding the search towards points on the lower bound of the cost/performance design space.

For performance estimation purposes we use a time-sampling technique (described in Section 4.3.4), which significantly speeds the simulation process. While this may not be highly accurate compared to full simulation, the fidelity is sufficient to make good incremental decisions guiding the search through the design space. To verify that our heuristic guides the search towards the pareto curve of the design space, we compare the exploration results with a full simulation of all the allocation and access pattern mapping alternatives for a large example. Indeed, as shown in Chapter 6, our algorithm finds the best cost/performance points in the design space, without requiring full simulation of the design space.

4.2.4   Experiments

We performed a set of experiments on a number of large multimedia and scientific applications to show the performance, cost and power tradeoffs generated by our approach.

Procedure Exploration
Input: Access Pattern Clusters, and the Memory Modules Library
Output: The Memory Architecture design points w/ best cost/perf ratios
begin
  Initialize the memory architecture to contain the initial_cache
  While cost of memory architecture < cost_constraint do
    While cost of memory architecture < cost_constraint and
      more allocations and mappings possible do
      For all access pattern clusters sharing a memory module
        Allocate a memory module and map the cluster to it
        Estimate cost of new memory architecture
        If (cost of new memory architecture > cost_constraint) continue
        Estimate performance of new memory architecture (time-sampling)
        Save current memory architecture alternative
        Undo memory module allocation and mapping
      end
      Choose the memory architecture with best cost/performance
      Perform full simulation of new design point
    end
    Double the cache size
  end
end

Figure 4.6. Exploration algorithm.

4.2.4.1   Experimental Setup

We simulated the design alternatives using our simulator based on the SIM-PRESS [MGDN01] memory model, and SHADE [CK93]. We assumed a processor based on the SUN SPARC [1], and we compiled the applications using gcc. We estimated the cost of the memory architectures (we assume the cost in equivalent basic gates) using figures generated by the Synopsys Design Compiler [Syn], and an SRAM cost estimation technique from [CWG+98].

We computed the average memory power consumption of each design point, using cache power figures from CACTI [WJ96]. For the main memory power consumption there is a lot of variation between the figures considered by different researchers [CWG+98, HWO97, VKI+00], depending on the main memory type, technology, and bus architecture. The ratio between the energy consumed by on-chip cache accesses and off-chip DRAM accesses varies between one and

---

[1] The choice of SPARC was based on the availability of SHADE and a profiling engine; however our approach is clearly applicable to any other (embedded) processor as well

two orders of magnitude [HWO97]. In order to keep our technique independent of such technology figures, we allow the designer to input the ratio R as:

$R = E\_main\_memory\_access/E\_cache\_access$

where E_cache_access is the energy for one cache access, and E_main_memory_access is the energy to bring in a full cache line. In our following power computations we assume a ratio R of 50, relative to the power consumption of an 8k 2-way set associative cache with line size of 16 bytes.

The use of multiple memory modules in parallel to service memory access requests from the CPU requires using multiplexers to route the data from these multiple sources. These multiplexers may increase the access time of the memory system, and if this is on the critical path of the clock cycle, it may lead to the increase of the clock cycle. We use access times from CACTI [WJ96] to compute the access time increases, and verify that the clock cycle is not affected.

Different cache configurations can be coupled with the memory modules explored, probing different areas of the design space. We present here our technique starting from an instance of such a cache configuration. A more detailed study for different cache configurations can be found in [GDN01c].

4.2.4.2 Results

Figure 4.7 presents the memory design space exploration of the access pattern customizations for the compress application. The compress benchmark exhibits a large variety of access patterns providing many customization opportunities. The x axis represents the cost (in number of basic gates), and the y axis represents the overall miss ratio (the miss ratio of the custom memory modules represents the number of accesses where the data is not ready when it is needed by the CPU, divided by the total number of accesses to that module).

The design points marked with a circle represent the memory architectures chosen during the exploration as promising alternatives, and fully simulated for accurate results. The design points marked only with an X represent the exploration attempts evaluated through fast time-sampling simulation, from which the best cost/gain tradeoff is chosen at each exploration step. For each such design we perform full simulation to determine accurate cost/performance/power figures.

The design point labeled 1 represents the initial memory architecture, containing an 8k 2-way associative cache. Our exploration algorithm evaluates the first set of customization alternatives, by trying to choose the best access pattern cluster to map to a custom memory module. The best performance gain for the incremental cost is generated by customizing the access pattern cluster containing a reference to the hashing table htab, which uses as an index in the array the access pattern reading the codetab array (the access pattern is htab[codetab[i]]). This new architecture is selected as the next design point in the exploration, labeled 2. After fully simulating the new memory architecture, we continue the exploration by evaluating the further possible customization opportunities, and selecting the best cost/performance ratio. In this way, we explore the memory architectures with most promising cost/performance tradeoffs, towards the lower bound of the design space.

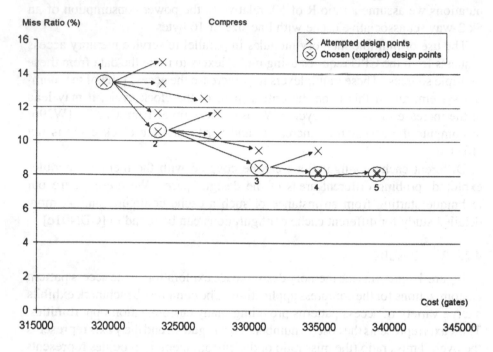

Figure 4.7. Miss ratio versus cost trade-off in Memory Design Space Exploration for Compress (SPEC95)

The miss ratio of the compress application varies between 13.42% for the initial cache-only architecture (for a cost of 319,634 gates), and 8.10% for a memory architecture where 3 access pattern clusters have been mapped to custom memory modules (for a cost of 334,864 gates). Based on a cost constraint (or alternatively on a performance requirement), the designer can choose the memory architecture which best matches the goals of the system.

In order to validate our space walking heuristic, and confirm that the chosen design points follow the pareto-curve-like trajectory in the design space, we compared the design points generated by our approach to pareto points generated by full simulation of the design space considering all the memory module allocations and access pattern cluster mappings for the compress example benchmark. Figure 4.8 shows the design space in terms of the estimated memory design cost (in number of basic gates), and the overall miss rate of the application. A design is on the pareto curve if there is no other design which is better in both cost and performance. The design points marked with an X represent the points explored by our heuristic. The points marked by a black dot, represent a full simulation of all allocation and mapping alternatives. The points on the lower bound of the design space are the most promising, exhibiting the best cost/performance tradeoffs. Our algorithm guides the search towards these design points, pruning the non-interesting points in the design space. Our exploration heuristic successfully finds the most promising designs, without fully simulating the whole design space: each fully simulated design on the lower bound (marked by a black dot) is covered by an explored design (marked by an X) [2]. This provides the designer the opportunity to choose the best cost/performance trade-off, without the expense of investigating the whole space.

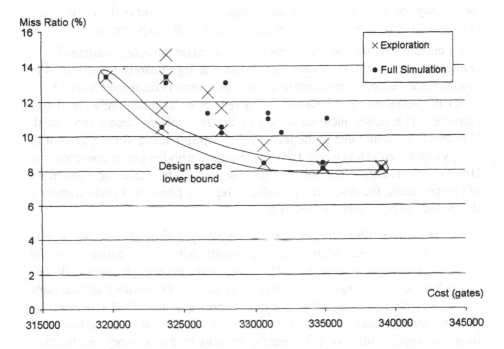

Figure 4.8. Exploration heuristic compared to simulation of all access pattern cluster mapping combinations for Compress

Table 4.1 presents the performance, cost and power results for a set of large, real-life benchmarks from the multimedia and scientific domains. The first column shows the application, and the second column represents the memory architectures explored for each such benchmark. The third column represents the cost of the memory architecture (in number of basic gates), the fourth column represents the miss ratio for each such design point, the fifth column shows the

---

[2]Not all exploration points (X) are covered by a full simulation point (black dot), since some of the exploration points represent estimations only

average memory latency (in cycles), and the last column presents the average memory power consumption, normalized to the initial cache-only architecture (represented by the first design point for each benchmark).

In Table 4.1 we present only the memory architectures with best cost/ performance characteristics, chosen during the exploration. The miss ratio shown in the fourth column represents the number of memory accesses when the data is not yet available in the cache or the custom memory modules when required by the CPU. The average memory latency shown in fifth column represents the average number of cycles the CPU has to wait for an access to the memory system. Due to the increased hit ratio, and to the fact that the custom memory modules require less latency to access the small FIFO containing the data than the latency required by the large cache tag, data array and cache control, the average memory latency varies significantly during the exploration.

By customizing the memory architecture based on the access patterns in the application, the memory system performance is significantly improved. For instance, for the compress benchmark, the miss ratio is decreased from 13.4% to 8.10%, representing a 39% miss ratio reduction for a relatively small cost increase. The power increase shown in the last column seems substantial. However it is mainly due to the increase in performance: a similar amount of energy consumed in less amount of time generates a higher power consumption. However, the energy variations are small. Moreover, by exploring a varied set of design points, the designer can tradeoff the cost, power and performance of the system, to best meet the design goals.

Vocoder is a multimedia benchmark exhibiting mainly stream-like regular access patterns, which behave well with small cache sizes. Since the initial cache of 1k has a small cost of 40,295 gates, there was enough space to double the cache size. The design points 1 through 4 represent the memory architectures containing the 1k cache, while the design points 5 through 8 represent the memory architectures containing the 2k cache. As expected, the performance increases significantly when increasing the cost of the memory architecture. However, a surprising result is that the power consumption of the memory system decreases when using the larger cache: even though the power consumed by the larger cache accesses increases, the main memory bandwidth decrease due to a lower miss ratio results in a significantly lower main memory power, which translates into a lower memory system power. Clearly, these types of results are difficult to determine by analysis alone, and require a systematic exploration approach to allow the designer to best trade off the different goals of the system.

The wide range of cost, performance, and power tradeoffs obtained are due to the aggressive use of the memory access pattern information, and customization of the memory architecture beyond the traditional cache architecture.

Benchmark	Design Point	Cost (gates)	Miss ratio (%)	Mem Latency (cycles)	Mem. Power (normalized)
Compress	1	319634	13.4200	28.56	1
	2	323521	10.5400	22.58	1.18
	3	330657	8.4500	18.42	1.36
	4	334864	8.1000	17.40	1.41
	5	339071	8.1000	17.35	1.41
li	1	319634	6.9800	15.82	1
	2	323841	4.6700	11.21	1.23
	3	332302	4.6200	11.01	1.24
	4	340763	4.6200	10.96	1.24
vocoder	1	40295	1.4600	4.90	1
	2	44502	1.3600	4.45	1.01
	3	48709	1.2600	4.16	1.02
	4	53765	1.2600	4.09	1.03
	5	80201	0.8100	3.61	0.68
	6	84408	0.7600	3.26	0.70
	7	88615	0.7400	3.13	0.70
	8	93671	0.7400	3.07	0.70

Table 4.1. Exploration results for our Access Pattern based Memory Customization algorithm.

4.2.5 Related Work

As outlined earlier in Section 2, there has been related work in four main domains: (I) High-level synthesis, (II) Computer Architecture, (III) Programmable embedded systems, and (IV) Disk file systems and databases.

(I) In the domain of High-Level Synthesis, custom synthesis of the memory architecture has been addressed for design of embedded ASICs. Catthoor et al. [CWG+98] address memory allocation, packing the data structures according to their size and bitwidth into memory modules from a library, to minimize the memory cost, and optimize port sharing. Wuytack et al. [WCdJ+96] present an approach to manage the memory bandwidth by increasing memory port utilization, through memory mapping and code reordering optimizations. Bakshi et al. [BG95] present a memory exploration approach, combining memory modules using different connectivity and port configurations for pipelined DSP systems. We complement this work by extracting and analyzing the prevailing accesses in the application in terms of access patterns, their relationships, similarities and interferences, and customize the memory architecture using memory modules from a library to generate a wide range of cost/performance/power tradeoffs in the context of programmable embedded systems.

(II) In the domain of Computer Architecture, [Jou90], [PK94] propose the use of hardware stream buffers to enhance the memory system performance.

Reconfigurable cache architectures have been proposed recently [VTG+99] to improve the cache behavior for general purpose processors, targeting a large set of applications. However, the extra control needed for adaptability and dynamic prediction of the access patterns while acceptable in general purpose computing where performance is the main target may result in a power overhead which is prohibitive in embedded systems, that are typically power constrained. Instead of using such dynamic prediction mechanisms, we statically target the local memory architecture to the data access patterns.

On a related front, Hummel et al. [HHN94] address the problem of memory disambiguation in the presence of dynamic data structures to improve the parallelization opportunities. Instead of using this information for memory disambiguation, we use a similar type of closed form description generated by standard compiler analysis to represent the access patterns, and guide the memory architecture customization.

(III) In the domain of programmable embedded systems, Kulkarni et al. [Kul01], Panda et al. [PDN99] have addressed customization of the memory architecture targeting different cache configurations, or alternatively using on-chip scratch pad SRAMs to store data with poor cache behavior. [GDN01a] presents an approach that customizes the cache architecture to match the locality needs of the access patterns in the application. However, this work only targets the cache architecture, and does not attempt to use custom memory modules to target the different access patterns.

(IV) In the domain of file systems and databases, there have been several approaches to use the file access patterns to improve the file system behavior. Parsons et al. [PUSS] present an approach allowing the application programmer to specify the file I/O parallel behavior using a set of templates which can be composed to form more complex access patterns. Patterson et al. [PGG+95] advocate the use of hints describing the access pattern (currently supporting sequential accesses and an explicit list of accesses) to select particular prefetching and caching policies in the file system.

The work we present differs significantly from all the related work in that we aggressively analyze, cluster and map memory access patterns to customized memory architectures; this allows the designer to trade-off performance and power against cost of the memory system.

4.3   Connectivity Architecture Exploration

In Section 4.2 we presented the exploration of the memory modules, based on the access patterns exhibited by the application, assuming a simple connectivity model. In this section we extend this work by performing connectivity exploration in conjunction with the memory modules exploration, to improve the behavior of the memory-connectivity system. There are two possible approaches to improving the memory system behavior: (a) a synthesis-oriented, optimization approach, where the result is a unique "best" solution, and (b) an exploration-based approach, where different memory system architectures are evaluated, and the most promising designs following a pareto-like shape are provided as the result, allowing the designer to further refine the choice, according to the goals of the system. We follow the second approach: we guide the design space search towards the pareto points in different design spaces (such as the cost/performance, and performance/power spaces), pruning the non-interesting designs early in the exploration process, and avoiding full simulation of the design space.

In Section 4.3.1 we present the flow of our approach. In Section 4.3.2 we use an example application to illustrate our exploration strategy, and in Section 4.3.3 we show the details of our Connectivity Exploration (ConEx) algorithm. We then present a set of experiments showing the cost, performance and power tradeoffs obtained by our coupled memory and connectivity exploration. In Section 4.3.5 we present related work in the area of connectivity and memory architecture exploration.

4.3.1   Our approach

Figure 4.9 shows the flow of our approach. The Connectivity Exploration (ConEx) approach is part of the MemorEx Memory System Exploration environment. Starting from the input application in C, our Access Pattern-based Memory Exploration (APEX) [GDN01b] algorithm first extracts the most active access patterns exhibited by the application data structures, and explores the memory module configurations to match the needs of these access patterns; however, it assumes a simple connectivity model. Our ConEx Connectivity Exploration approach starts from this set of selected memory modules configurations generated by APEX, and explores the most promising connectivity architectures, which best match the performance, cost and power goals of the system. Since the complete design space is very large, and evaluating all possible combinations in general is intractable, at each stage we prune out the non-interesting design configurations, and consider for further exploration only the points which follow a pareto-like curve shape in the design space.

In order to test that our exploration strategy successfully follows the pareto-like curve shape, we compare our pruning with a full space simulation for two large applications.

Starting from a memory architecture containing a set of memory modules, we map the communication channels between these modules, the off-chip memory and the CPU to connectivity modules from a connectivity IP library. Figure 4.10 (a) shows the connectivity architecture template for an example memory architecture, containing a cache, a stream buffer, an on-chip SRAM, and an off-chip DRAM. The communication channels between the on-chip memory modules, the off-chip memory modules and the CPU can be implemented in many ways.

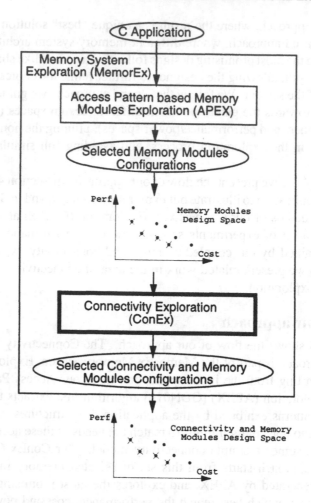

Figure 4.9.   The flow of our Exploration Approach.

One naive implementation is where each communication channel is mapped to one connectivity module from the library. However, while this solution may generate good performance, in general the cost is prohibitive. Instead, we cluster the communication channels into groups based on their bandwidth requirement, and map each such cluster to connectivity modules. Figure 4.10 (b) shows an example connectivity architecture implementing the communication channels, containing two on-chip busses, a dedicated connection, and an off-chip bus.

4.3.2   Illustrative example

We use the compress benchmark (SPEC95) to illustrate the cost, performance, and power trade-offs generated by our connectivity exploration ap-

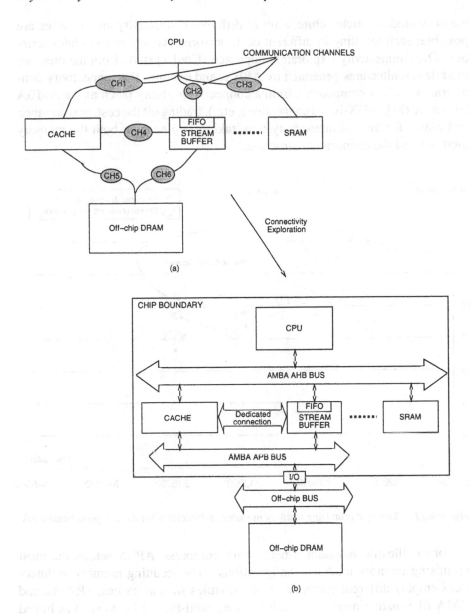

Figure 4.10. (a) The Connectivity Architecture Template and (b) An Example Connectivity Architecture.

proach. The benchmark contains a varied set of data structures and access patterns, presenting interesting opportunities for customizing the memory architecture and connectivity.

First we perform Access Pattern based Memory Exploration (APEX) [GDN01b], to determine a set of promising memory modules architectures. For each such memory modules architecture, a set of different connectivity architectures are possible, each resulting in different cost, performance and power characteristics. Our Connectivity Exploration approach (ConEx) starts from the memory modules architectures generated by APEX, and explores the connectivity configurations, using components from a connectivity library (such as the AMBA busses [ARM], MUX-based connections, etc.), trading off the cost, performance and power for the full memory system, taking into account both the memory modules and the connectivity structure.

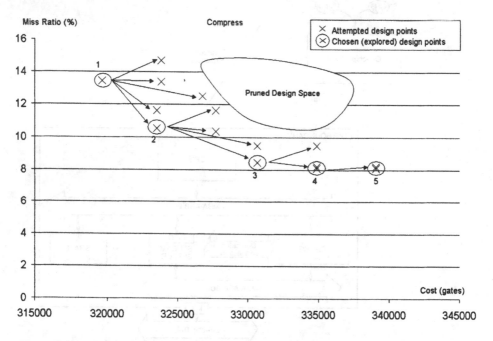

Figure 4.11. The most promising memory modules architectures for the compress benchmark.

For the illustrative example benchmark compress, APEX selects the most promising memory modules configurations. The resulting memory architectures employ different combinations of modules such as caches, SRAMs, and DMA-like custom memory modules storing well-behaved data such as linked lists, arrays of pointers, streams, etc. [GDN01b]. Figure 4.11 shows the memory modules architectures explored by APEX for the compress example (this figure is similar to Figure 4.7; we duplicate it for convenience, indicating also the pruned design space). The X axis represents the cost of the memory modules in basic gates, and the Y axis represents the overall miss ratio (we assume that accesses to on-chip memory such as the cache or SRAM are hits, and accesses to off-chip memory are misses). APEX prunes the non-interesting designs, on the inside of the pareto curve, choosing only the most promising cost/performance architectures for further exploration. The points labeled 1 through 5 represent the selected memory modules designs, which will be used as the starting point for the connectivity exploration.

Each such selected memory architecture may contain multiple memory modules with different characteristics, and communication requirements. For each such architecture, different connectivity structures, with varied combinations of connectivity modules from the library may be used. For instance, the memory modules architecture labeled with 3 in Figure 4.11 contains a cache, a memory module for stream accesses, a memory module for self-indirect array references (as explained in Section 4.2) and an off-chip DRAM. When using dedicated or MUX-based connections from the CPU to the memory modules, the latency of the accesses is small, at the expense of longer connection wires. Alternatively, when using a bus-based connection, such as the AMBA System bus (ASB) [ARM], the wire length decreases, at the expense of increased latency due to the more complex arbitration needed. Similarly, when using wider busses, with pipelined or split transaction accesses, such as the AMBA High-performance bus (AHB) [ARM], the wiring and bus controller area increases further. Moreover, all these considerations impact the energy footprint of the system. For instance, longer connection wires generate larger capacitances, which may lead to increased power consumption.

Figure 4.12 shows the ConEx connectivity exploration for the compress benchmark. The X axis represents the cost of the memory and connectivity system. The Y axis represents the average memory latency, including the latency due to the memory modules, as well as the latency due to the connectivity. The average memory latency is reduced from 10.6 cycles to 6.7 cycles, representing a 36% improvement [3], while trading off the cost of the connectivity and memory modules.

Alternatively, for energy-aware designs, similar tradeoffs are obtained in the cost/power or the performance/power design spaces (the energy consumption tradeoffs are presented in the Chapter 6). In this manner we can customize the connectivity architecture, thus substantially improving the memory and connectivity system behavior, and allowing the designer to trade off the different goals of the system.

4.3.3 Connectivity Exploration Algorithm

Our Connectivity Exploration (ConEx) algorithm is a heuristic method to evaluate a wide range of connectivity architectures, using components from a

---

[3]For clarity in the figures, we did not include the uninteresting designs exhibiting very bad performance (many times worse than the best designs). While those designs would increase even further the performance variation, in general they are not useful.

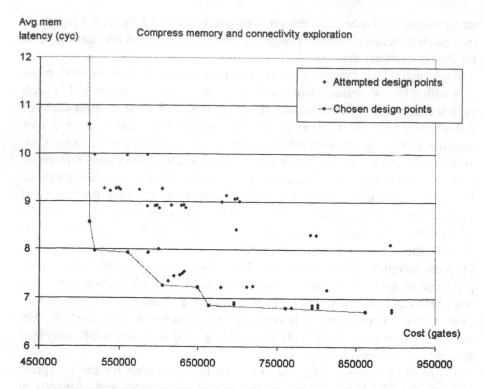

Figure 4.12. The connectivity architecture exploration for the compress benchmark.

connectivity IP library, and selecting the most promising architectures, which best trade-off the connectivity cost, performance and power.

Figure 4.13 shows our Connectivity Exploration algorithm. The input to our ConEx algorithm is the application in C, a set of selected memory modules architectures (generated by the APEX exploration step in Section 4.2), and the connectivity library. Our algorithm generates as output the set of most promising connectivity/memory modules architectures, in terms of cost, performance and power.

For each memory modules architecture selected in the APEX memory modules exploration stage from Section 4.2, multiple connectivity implementations are possible. Starting from these memory modules architectures, we explore the connectivity configurations by taking into account the behavior of the complete memory and connectivity system, allowing the designer to tradeoff the cost, performance and power of the design. The ConEx algorithm proceeds in two phases: (I) Evaluate connectivity configurations (II) Select most promising designs.

(I) **Evaluate connectivity configurations**. For each memory architecture selected from the previous APEX Memory Modules Exploration phase (Section 4.2), we evaluate different connectivity architecture templates and connectivity allocations using components from the connectivity IP library. We estimate the cost, performance and power of each such connectivity architecture, and perform an initial selection of the most promising design points for further evaluation.

We start by profiling the bandwidth requirement between the memory modules and CPU for each memory modules architecture selected from APEX, and constructing a Bandwidth Requirement Graph (BRG). The Bandwidth Requirement Graph (BRG) represents the bandwidth requirements of the application for the given memory modules architecture. The nodes in the BRG represent the memory and processing cores in the system (such as the caches, on-chip SRAMs, DMAs, off-chip DRAMs, the CPU, etc.), and the arcs represent the channels of communication between these modules. The BRG arcs are labeled with the average bandwidth requirement between the two modules.

Each arc in the BRG has to be implemented by a connectivity component from the connectivity library. One possible connectivity architecture is where each arc in the BRG is assigned to a different component from the connectivity library. However, this naive implementation may result in excessively high cost, since it does not try to share the connectivity components. In order to allow different communication channels to share the same connectivity module, we hierarchically cluster the BRG arcs into logical connections, based on the bandwidth requirement of each channel. We first group the channels with the lowest bandwidth requirements into logical connections. We label each such cluster with the cumulative bandwidth of the individual channels, and continue the hierarchical clustering. For each such clustering level, we then explore all feasible assignments of the clusters to connectivity components from the library, and estimate the cost, performance, and power of the memory and connectivity system.

(II) **Select most promising designs.** In the second phase of our algorithm, for each memory and connectivity architecture selected from Phase I we perform full simulation to determine accurate performance and power metrics. We then select the best combined memory and connectivity candidates from the simulated architectures.

While in Phase I we selected separately for each memory module architecture the best connectivity configurations, in Phase II we combine the selected designs and choose the best overall architectures, in terms of both the memory module and connectivity configuration.

The different design points present different cost, performance and power characteristics. In general, these three optimization goals are incompatible. For instance, when optimizing for performance, the designer has to give up either cost, or power. Typically, the pareto points in the cost/performance space have a poor power behavior, while the pareto points in the performance/power space will incur a large cost. We select the most promising architectures using three

Procedure ConnectivityExploration(Memory Modules Architecture *mem_arch*)
Input: The C Application and the Memory Modules Architecture *mem_arch*,
  the Connectivity Library
Output: The most promising Connectivity Design Points
begin
  Profile the Memory Modules Architecture *mem_arch*
  Construct the Bandwidth Requirement Graph (BRG)
  Allocate each arc in the BRG to a logical connection cluster
  connect_design_points = ϕ
  do{
    if number_of_logical_connections $\leq$ max_cost_constraint
    Allocate the logical connections to physical connections
      from the Connectivity Library
    Estimate the Cost, Performance and Power of connecitivity architecture
    Add this connectivity architecture to connect_design_points
  Merge the two logical connection clusters with lowest bandwidth requirement
    hierarchycally into a larger cluster
  }while(more clusters can be merged)
  return connect_design_points;
end

Algorithm ConEx
Input: C Application, Selected Memory Modules Architectures
Output: The Combined Memory Modules and Connectivity Design Points
  w/ best cost/performance/power trade-offs
begin
  Phase I:
    combined_design_points = ϕ
    For each selected memory module architecture *mem_arch*
    connect_design_points = ConnectivityExploration(*mem_arch*)
    Select the local most promising connectivity desing points from
      connect_design_points
    Add selected design points to combined_design_points
  Phase II:
    Simulate the design points from combined_design_points
    Select the global most promising combined memory modules and connectivity
      design points from combined_design_points
end

Figure 4.13.   Connectivity Exploration algorithm.

scenarios: (a) In a power-constrained scenario, where the energy consumption has to be less then a threshold value, we determine the cost/performance pareto points, to optimize for cost and performance, while keeping the power less then the constraint, (b) In a cost-constrained scenario, we compute the performance/power pareto points, and (c) In a performance-constrained scenario, we compute the pareto points in the cost-power space, optimizing for cost and power, while keeping the performance within the requirements.

(a) In the power-constrained scenario, we first determine the pareto points in the cost-performance space. Recall that a design is on the pareto curve if there is no other design which is better in both cost and performance. We then collect the energy consumption information for the selected designs. The points on the cost-performance pareto curve may not be optimal from the the energy consumption perspective. From the selected cost-performance pareto points we choose only the ones which satisfy the energy consumption constraint. The designer can then tradeoff the cost and performance of the system to best match the design goals.

(b) In the cost-constrained scenario, we start by determining the pareto points in the performance-power space, and use the system cost as a constraint. Conversely, the pareto points in the performance-power space are in general not optimal from the cost perspective.

(c) When using the performance as a constraint, we determine the cost-power pareto points.

For performance and power estimation purposes we use a time-sampling technique [KHW91] (see Section 4.3.4 for a description of the time-sampling technique), which significantly speeds the simulation process. While this may not be highly accurate compared to full simulation, the fidelity is sufficient to make good incremental decisions guiding the search through the design space. To verify that our heuristic guides the search towards the pareto curve of the design space, we compare the exploration results with a full simulation of all the memory and connectivity mapping alternatives for two large examples. Indeed, as shown in Chapter 6, our algorithm successfully finds the best points in the design space, without requiring full simulation of the design space.

4.3.3.1   Cost, performance, and power models

We present in the following the cost, performance and power models used during our memory modules and connectivity exploration approach.

The cost of the chip is composed of two parts: the cost of the cores (such as CPU cores, memories and controllers), and the cost of the connectivity wiring. We use the method employed in [CZY$^+$99] to compute the total chip area: since the wiring area and the cores area can be proportionally important, we use two factors α and β tuned so that the overall wire and core areas are balanced [CZY$^+$99]:

$Chip\_area = \alpha * Wire\_area + \beta * Cores\_area$

where the Wire_area is the area used by the connectivity wires, and the Cores_area is the area of the memory modules, memory controllers, CPU cores, and bus controllers.

For the connectivity cost we consider the wire length and bitwidth of the busses, and the complexity of the bus controller. We estimate the wire length to half the perimeter of the modules connected by that wire [CWG+98]:

$$Conn\_length = \Sigma(2 * \sqrt{module\_area})$$

where the sum is over all the modules connected by that connectivity component, and the module_area is the area of each such module. The area of this connectivity is then:

$$Conn\_area = \alpha * Conn\_length * Conn\_bitwidth + \beta * Controller\_area$$

where Conn_bitwidth is the connectivity bitwidth, Controller_area is the area of the connectivity controller, and α and β are the two scaling factors determined as mentioned above.

We determine the cost of the on-chip cache using the cost estimation techniques presented in [CWG+98]. For the cost of the custom memory modules explored in the previous stage of our DSE approach [GDN01c] we use figures from the Synopsys Design Compiler [Syn]. For the CPU area (used to compute the wire lengths for the wires connecting the memory modules to the CPU), we use figures reported for the LEON SPARC gate count in 0.25um [Gai].

Since the area of the off-chip memory is not as important as for the on-chip components, we do not consider the off-chip memory area into our cost function.

We compute the performance of the memory system by generating a memory simulator for the specific memory modules and connectivity architectures [MGDN01]. We describe the timings and pipelining of the memory and connectivity components using Reservation Tables, as presented in [GDN00, GHDN99] and explained in Chapter 5. Busses may support continuous or split transactions, with different levels of pipelining. These features are also captured using the Reservation Tables model, augmented with explicit information on the presence of split or continuous transactions.

The memory system energy consumption is composed of two parts: the connectivity energy consumption, and the memory modules energy consumption.

We estimate the energy consumption of the connectivity components based on the power estimation technique presented in [CWG+98]:

$$Econn/access = 1/2 * Bus\_bitwidth * Atoggle * Fclock * (Cdriver + Cload) * Vdd^2 * access\_latency$$

where Econn/access is the energy per access consumed by the connectivity module, Bus_bitwidth is the bitwidth of the bus, Atoggle is the probability that a bit line toggles between two consecutive transfers, Fclock is the clock frequency, and Cdriver and Cload are the capacitance of the buffer that drives the connectivity, and the total load capacitance of the wires.

We compute the load capacitance of the on-chip interconnect as:

$$Cload = Lconn * Cmm,$$

where Lconn is the length of the connectivity (computed as described above), and Cmm is the capacitance per mm of the wires. We assume a capacitance of 0.19pF/mm for 0.25um technology [CWG$^+$98]. While this figure is for the first metal layer, the capacitance values for the different layers are not dramatically different [CWG$^+$98].

For the driver capacitance we use the approach presented in [LS94]. Assuming that the size ratio in the inverter chain is 4, and the inverter that drives the load has a capacitance of 1/4 of its load, the total capacitance of the buffer is about 30% of the total load. The total capacitance ratio of the inverter chain is 1/4 + 1/16 + 1/64 + ... = 0.3:

$Cdriver = Cload * 0.3,$

We compute the cache energy consumption per access using Cacti [WJ96]. We determine the energy consumed by off-chip accesses, including the off-chip DRAM power, I/O pins, and assuming 30mm off-chip bus length [CWG$^+$98]. For the off-chip connectivity, we use the capacitance figures presented in [CWG$^+$98]: a typical off-chip bus capacitance is 0.1pF/mm (we assume a 30mm off-chip bus), and the bus driver capacitance is 5pF. The chip I/O pins capacitance varies between 1pF and 10pF depending on the packaging type (we assume a capacitance of 5pF for the I/O pins).

For the off-chip DRAM energy consumption there is a lot of variation between the figures considered by different researchers [CWG$^+$98, HWO97, VKI$^+$00], depending on the main memory type, technology. The ratio between the energy consumed by on-chip cache accesses and off-chip DRAM accesses varies significantly ([CWG$^+$98] reports a ratio of 3 to 5 for accesses of same size, and [HWO97] reports a ratio between one and two orders of magnitude; however, it is not clear weather this includes the connectivity energy). In order to keep our technique independent of such technology figures, and allow the designer to determine the relative importance of these factors, we define a ratio R:

$R = Emain\_memory\_access/Ecache\_access$

where E_main_memory_access and Ecache_access are the energy consumed per access by the main memory and the cache for accesses of the same size. We assume a ratio of 5, compared to an 8k 2-way set associative cache.

We assume the energy consumed by the custom memory controllers presented in [GDN01c] to be similar to the energy consumed by the cache controller.

4.3.3.2 Coupled Memory/Connectivity Exploration strategy

The quality of the final selected memory-connectivity architectures in different spaces, such as cost/performance, or cost/power spaces, depends on the quality of the initial memory modules architectures selected as starting points for the connectivity exploration. The memory modules architecture selection has to be driven by the same metric as the connectivity architecture selection. For instance, when cost and performance are important, we guide the search towards the cost/performance pareto points both in the early APEX memory modules exploration, as well as in the ConEx connectivity exploration, and use power as a constraint. Alternatively, when cost and power are important, we use cost/power as the metric to guide both the APEX and the ConEx explorations. For this, we modified the APEX [GDN01b] algorithm to use cost/power as the exploration driver, to determine the cost/power pareto points for the memory modules architectures. We then use these architectures as the starting point for the connectivity exploration.

In order to verify the sensitivity of the exploration on the memory modules architectures used as starting point, we compare three exploration strategies, using different sets of starting memory modules architectures: (I) Pruned exploration, where we select the most promising memory modules and connectivity architectures, and perform full simulation to determine the pareto curve without fully exploring the design space, (II) Neighborhood exploration, where we expand the design space by including also the points in the neighborhood of the architectures selected in the Pruned approach, and (III) Full space exploration, the naive approach where we fully simulate the design space, and compute the pareto curve.

(I) In the Pruned exploration approach we start by selecting the most promising memory modules configurations, and use them as input for the connectivity exploration phase. For each such memory module architecture, we then explore different connectivity designs, estimating the cost, performance and energy consumption, and selecting at each step the best cost, performance and power tradeoffs. We then simulate only the selected designs and determine the pareto points from this reduced set of alternatives, in the hope that we find the overall pareto architectures, without fully simulating the design space.

(II) In order to increase the chances of finding the designs on the actual pareto curve, we expand the explored design space by including the memory modules architectures in the neighborhood of the selected designs. In general, this leads to more points in the neighborhood of the pareto curve being evaluated, and possibly selected.

(III) We compare our Pruned and Neighborhood exploration approaches to the brute force approach, where we fully simulate the design space and fully determine the pareto curve. Clearly, performing full simulation of the design space is very time consuming and often intractable. We use the naive Full space exploration approach only to verify that our Pruned and Neighborhood exploration strategies successfully find the pareto curve designs points, while significantly reducing the computation time.

Clearly, by intelligently exploring the memory modules and connectivity architectures using components from a library, it is possible to explore a wide range of memory system architectures, with varied cost, performance and power characteristics, allowing the designer to best tradeoff the different goals of the system. We successfully find the most promising designs following the pareto-like curve, without fully simulating the design space.

4.3.4 Experiments

We present Design Space Exploration (DSE) results of the combined memory modules and connectivity architecture for a set of large realistic benchmarks to show the performance, cost and power tradeoffs generated by our approach. The memory modules architectures selected by our memory modules exploration presented in Section 4.2 have been used as starting point for our connectivity exploration. We present here the experimental results taking into account the cost, performance and power for the full memory system, including both the memory and the connectivity architecture.

We attempt to find the pareto curve designs without simulating the full design space. Our exploration algorithm guides the search towards the points on the pareto curve of the design space, pruning out the non-interesting designs (for instance, assuming a two dimensional cost-performance design space, a design is on the pareto curve, if there is no other design which is better in terms of both cost and performance).

In order to verify that our Design Space Exploration (DSE) approach successfully finds the points on the pareto curve, we compare the exploration algorithm results with the actual pareto curve obtained by fully simulating the design space.

4.3.4.1 Experimental Setup

We simulated the design alternatives using our simulator based on the SIM-PRESS [MGDN01] memory model, and SHADE [CK93]. We assumed a processor based on the SUN SPARC [4], and we compiled the applications using gcc.

We used a set of large real-life multimedia and scientific benchmarks. Compress and Li are from SPEC95, and Vocoder is a GSM voice encoding application.

We use time-sampling estimation to guide the walk through the design space, pruning out the designs which are not interesting. We then use full simulation for the most promising designs, to further refine the tradeoff choices. The time sampling alternates "on-sampling" and "off-sampling" periods, assuming a ratio of 1/9 between the on and off time intervals.

---

[4]The choice of SPARC was based on the availability of SHADE and a profiling engine; however our approach is clearly applicable to any other embedded processor as well

The time-sampling estimation does not have a very good absolute accuracy compared to full simulation. However, we use it only for relative incremental decisions to guide the design space search, and the estimation fidelity is sufficient to make good pruning decisions. The estimation error is due to the cold start problem [KHW91]: in large caches, the data accessed during a long off-sampling period is not updated in the cache. As a result, the misses which occur at the start of the next "on-sampling" period may not be real misses, but rather due to the fact that the data has not been brought in the cache during the previous "off-sampling" period. As a result the time sampling constantly over-estimates the miss ratio [KHW91], and the fidelity of the estimation is sufficient for the relative comparisons needed in the design space search. Moreover, the cold-start problem is especially important in large multi-megabyte caches, such as the ones used in [KHW91] (e.g., for a 1M cache, time-sampling overestimates the miss ratio with roughly 20% for most benchmarks), and increases with the size of the cache. However, in embedded systems, due to space constraints, the on-chip caches are significantly smaller. Furthermore, the alternative to time-sampling is set-sampling, performing sampling along the sets of the cache [KHW91]. While set-sampling generates better absolute accuracy, it is not suited in the presence of time-dependent techniques such as data prefetching, where time-sampling is required [KHW91].

4.3.4.2   Results

We performed two sets of experiments: (I) Using cost/ performance to drive the memory and connectivity exploration, and (II) Using cost/power to drive the exploration. In each such set of experiments we present the effect of the exploration in all the three dimensions (cost, performance and power).

(I) Figure 4.14 shows the cost/performance tradeoff for the connectivity exploration of the compress benchmark. The X axis represents the cost of the memory and connectivity architecture, and the Y axis represents the average memory latency including both the memory and connectivity latencies (e.g., due to the cache misses, bus multiplexing, or bus conflicts).

In this experiment we used cost/ performance to drive the selection algorithms both during the memory modules exploration, and during the connectivity exploration. The dots represent the attempted connectivity and memory designs. The line connecting the squares represents the designs on the cost/ performance pareto. However, the designs which have best cost/performance behavior, do not necessarily have good power behavior. The line connecting the triangles represents the designs in the cost/performance space which are on the performance/power pareto curve. While the cost/performance and the performance/power pareto curves do not coincide, they do have a point in common. However, this point has a very large cost. In general, when trading off cost, performance and power, the designer has to give up one of the goals in order to optimize the other two. For instance, if the designer wants to optimize performance and power, it will come at the expense of higher cost.

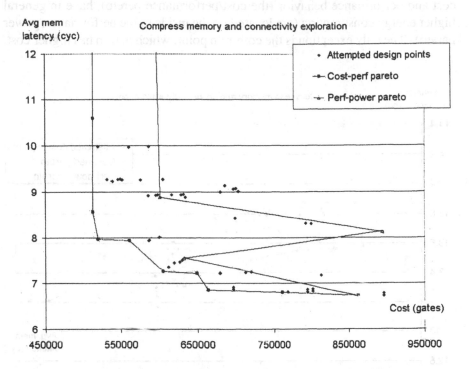

Figure 4.14. Cost/perf vs perf/power paretos in the cost/perf space for Compress.

Figure 4.15 shows the performance/power tradeoffs for the connectivity exploration of the compress benchmark, using cost/performance to drive the selection of the starting memory modules. The X axis represents the average memory latency, including both the memory and connectivity components. The Y axis represents the average energy per access consumed by the memory and connectivity system. We use energy instead of average power consumption, to separate out the impact of the actual energy consumed from the variations in performance. Variations in total latency may give a false indication of the power behavior: for instance, when the performance decreases, the average power may decrease due to the longer latency, but the total energy consumed may be the same.

The line connecting the squares represents the cost/performance pareto points in the performance/power space. The line connecting the triangles, shows the performance/power pareto points. Again, the best performance/power points do not necessarily have also low cost. The cost/performance pareto, and the performance/power pareto do not coincide in the performance/power space.

When trading off the three goals of the system, the designer has to give up one of the dimensions, in order to optimize the other two. The designs which have good cost and performance behavior (the cost/performance pareto), have in general higher energy consumption (are located on the inside of the performance/power pareto). The only exception is the common point, which in turn has higher cost.

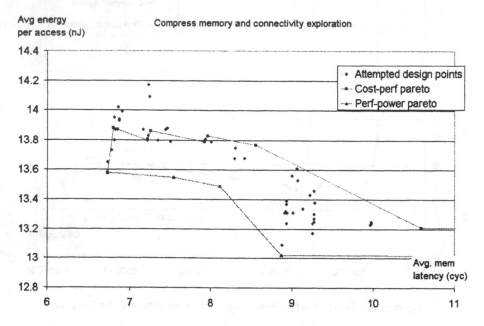

Figure 4.15.   Cost/perf vs perf/power paretos in the perf/power space for Compress.

(II) Figure 4.16 and Figure 4.17 show the cost/performance and performance/ power spaces for the exploration results for compress, using cost/power to drive the memory and connectivity exploration.

In Figure 4.16, the line connecting the squares represents the cost/performance pareto obtained by the experiments where cost/power was used to guide the exploration, and the line connecting the stars represents the cost/performance pareto in the case where cost/performance was used throughout the exploration as the driver. As expected, the best cost/performance points obtained during the cost/performance exploration are better in terms of cost and performance than the ones obtained during the cost/power exploration.

In Figure 4.17, the line connecting the triangles represents the performance/ power pareto for the cost/performance exploration, while the line connecting the stars represents the performance/power pareto for the cost/power exploration. As expected, when using cost/power to drive the early memory modules exploration (APEX), the overall energy figures are better.

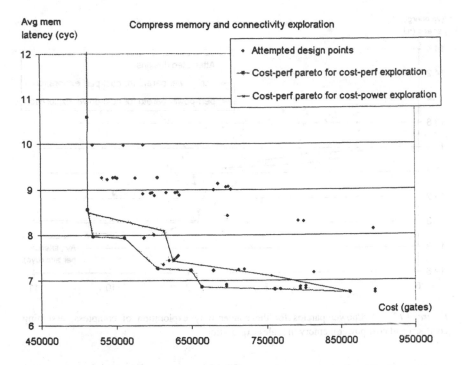

Figure 4.16. Cost/perf paretos for the connectivity exploration of compress, assuming cost/perf and cost/power memory modules exploration.

Similarly, the cost/power space representation of the cost/power exploration yields better results in terms of cost and power then the cost/performance exploration.

We performed similar experiments on the Vocoder and Li benchmarks. Figure 4.18 and Figure 4.19 show the comparison between the cost/performance and the performance/power paretos for the connectivity exploration, assuming that the previous phase of memory modules exploration is driven by cost and power.

Figure 4.20 and Figure 4.21 show the comparison between the cost/performance and the performance/power paretos for the connectivity exploration for Vocoder, while Figure 4.22 and Figure 4.23 show the comparison between the cost/performance and the performance/power paretos for the connectivity exploration for Li.

Figure 4.24 shows the analysis of the cost/performance pareto-like points for the compress benchmark. The design points *a* through *k* represent the most promising selected memory-connectivity architectures. Architectures *a* and *b* represent two instances of a traditional cache-only memory configuration, using the AMBA AHB split transaction bus, and a dedicated connection. The architectures *c* through *k* represent different instances of novel memory and

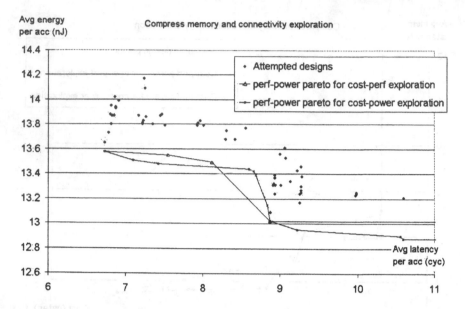

Figure 4.17. Perf/power paretos for the connectivity exploration of compress, assuming cost/perf and cost/power memory modules exploration.

connectivity architectures, employing SRAMs to store data which is accessed often, DMA-like memory modules to bring in predictable, well-known data structures (such as lists) closer to the CPU, and stream buffers for stream-based accesses. Architecture c contains a linked-list DMA-like memory module, implementing an self-indirect data structure, using a MUX-based connection. This architecture generates a roughly 10% performance improvement for a small cost increase, over the best traditional cache architecture (b). The architecture d represents the same memory configuration as c, but with a connectivity containing both a MUX-based structure and an AMBA APB bus. Similarly, architectures e through k make use of additional linked-list DMAs, stream buffers, and SRAMs, with MUX-based, AMBA AHB, ASB and APB connections. Architecture g generates a roughly 26% performance improvement over the best traditional cache architecture (b), for a roughly 30% memory cost increase. Architecture k shows the best performance improvement, of roughly 30% over the best traditional cache architecture, for a larger cost increase. Clearly, our memory-connectivity exploration approach generates a significant performance improvement for varied cost configurations, allowing the designer to select the most promising designs, according to the available chip space and performance requirements.

Figure 4.25 represents the analysis of the cost/performance pareto-like architectures for the vocoder benchmark. The architectures a and b represent the

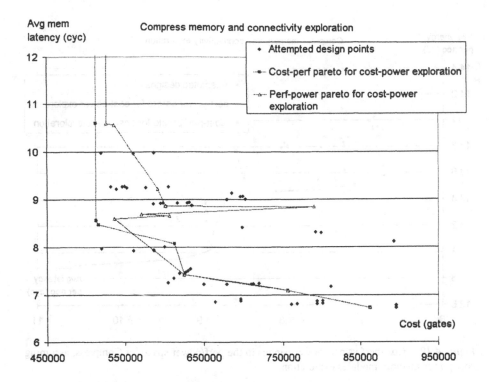

Figure 4.18. Cost/perf vs perf/power paretos in the cost/perf space for Compress, assuming cost-power memory modules exploration.

traditional cache architectures with AMBA AHB and dedicated connections. The architecture *c* containing the traditional cache and a stream buffer generates a 5% performance improvement over the best traditional cache architecture (*b*) for a roughly 3% cost increase. Due to the fact that the vocoder application is less memory intensive, containing mainly stream-based accesses, which behave well on cache architectures, the performance variation is less significant then in the other benchmarks. However, this illustrates the application-dependent nature of the memory and bandwidth requirements of embedded systems, prompting the need for early memory and connectivity exploration. Clearly, without such an exploration framework it would be difficult to determine through analysis alone the number, amount and type of memory modules required to match the given performance, energy and cost criteria.

Figure 4.26 represents the analysis of the cost/performance pareto-like architectures for the Li benchmark. The memory-connectivity architectures containing novel memory modules, such as linked-list DMAs implementing self-indirect accesses, and stream buffers, connected through AMBA AHB, ASB

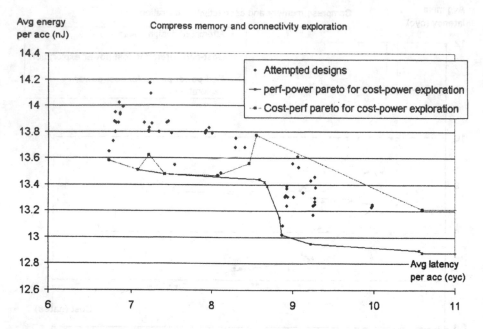

Figure 4.19. Cost/perf vs perf/power paretos in the perf/power space for Compress, assuming cost-power memory modules exploration.

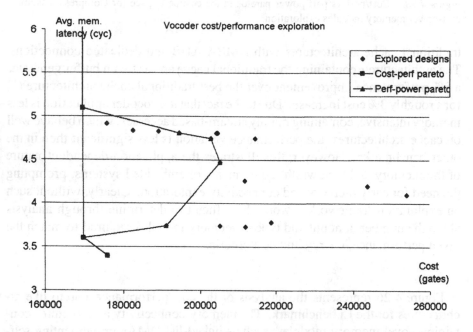

Figure 4.20. Cost/perf vs perf/power paretos in the cost/perf space for Vocoder.

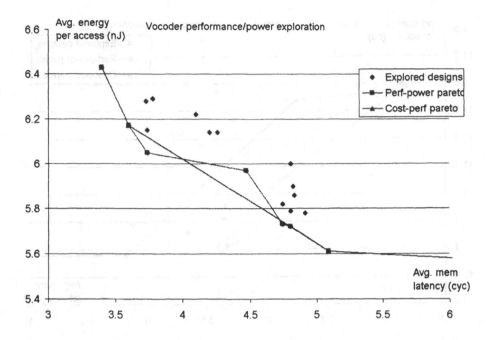

Figure 4.21. Cost/perf vs perf/power paretos in the perf/power space for Vocoder.

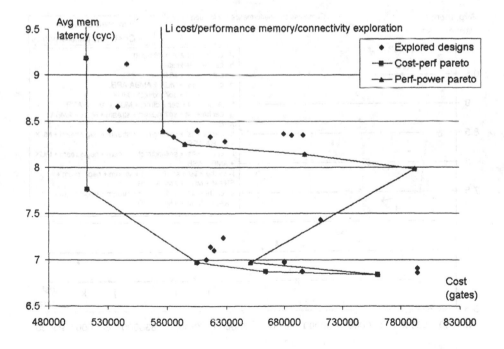

Figure 4.22. Cost/perf vs perf/power paretos in the cost/perf space for Li.

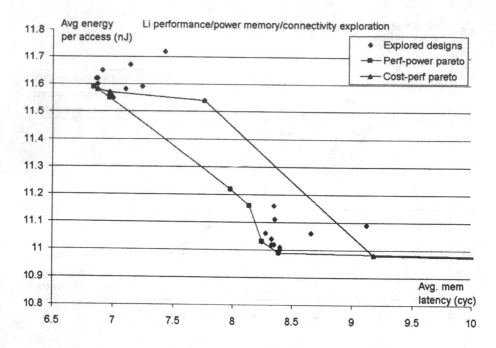

Figure 4.23.   Cost/perf vs perf/power paretos in the perf/power space for Li.

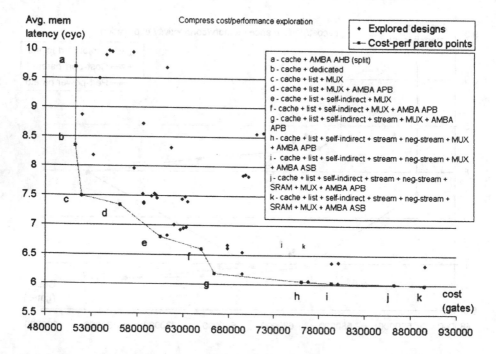

Figure 4.24.   Analysis of the cost/perf pareto architectures for the compress benchmark.

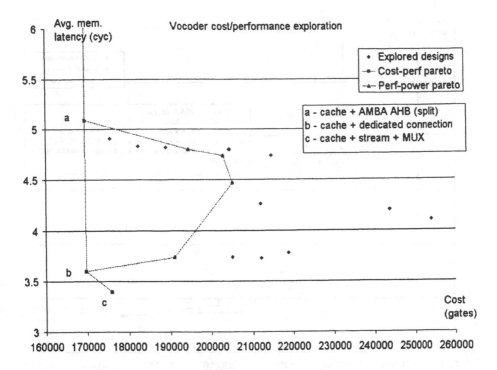

Figure 4.25. Analysis of the cost/perf pareto architectures for the vocoder benchmark.

and APB busses, generate significant performance variations, allowing the designer to best match the requirements of the system.

In the following we present the exploration results for the compress, Li, and vocoder benchmarks. We show only the selected most promising cost/performance designs, in terms of their cost (in basic gates), average memory latency, and average energy consumption per access. In Table 4.2 the first column shows the benchmarks, the second, third and fourth columns show the cost, average memory latency and energy consumption for the selected design simulations. The simulation results show significant performance improvement for varied cost and power characteristics of the designs, for all the benchmarks. For instance, when using different memory and connectivity configurations, the performance of the compress and Li benchmarks varies by an order of magnitude. The energy consumption of these benchmarks does not vary significantly, due to the fact that the connectivity consumes a small amount of power compared to the memory modules.

Table 4.3 presents the coverage of the pareto points obtained by our memory modules and connectivity exploration approach. Column 1 shows the benchmark, and Column 2 shows the category: Time represents the total computation time required for the exploration, Coverage shows the percentage of the points on the pareto curve actually found by the exploration. Average distance shows

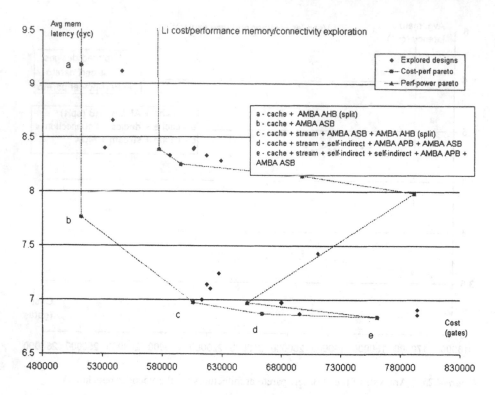

Figure 4.26.   Analysis of the cost/perf pareto architectures for the Li benchmark.

the average percentile deviation in terms of cost, performance and energy consumption, between the pareto points which have not been covered, and the closest exploration point which approximates them. Column 3 represents the results for the Pruned exploration approach, where only the most promising design points from the memory modules exploration are considered for connectivity space exploration. Column 4 shows the Neighborhood exploration results, where the design points in the neighborhood of the selected points are also included in the exploration, and the last Column shows the results for the brute-force full space exploration, where all the design points in the exploration space are fully simulated, and the pareto curve is fully determined.

Note that the average distance in Table 4.3 is small, which indicates that even though a design point on the pareto curve was not found, another design with very close characteristics (cost, performance, power) was simulated (i.e., there are no significant gaps in the coverage of the pareto curve).

In the Pruned approach during each Design Space Exploration phase we select for further exploration only the most promising architectures, in the hope that we will find the pareto curve designs without fully simulating the design space. Neighborhood exploration expands the design space explored, by in-

Benchark	Cost [gates]	Avg mem latency [cycles]	Avg energy [nJ]
Compress	480775	69.66	13.24
	512232	62.76	13.52
	512332	9.69	13.80
	512532	8.35	14.36
	519388	7.49	14.44
	561112	7.34	14.39
	604941	6.80	14.47
	649849	6.60	14.39
	664029	6.19	14.46
	760543	6.05	14.47
	793971	6.03	14.54
	862176	6.01	14.31
	895604	5.99	14.38
li	480775	57.59	10.42
	494992	57.48	10.43
	512232	50.29	10.70
	512332	9.18	10.98
	512532	7.76	11.54
	605767	6.97	11.57
	664029	6.87	11.58
	760543	6.84	11.59
vocoder	156806	16.37	5.05
	169370	13.28	5.33
	169481	5.09	5.61
	169703	3.60	6.17
	175865	3.40	6.43

Table 4.2. Selected cost/performance designs for the connectivity exploration.

cluding also the points in the neighborhood of the points selected by the Pruned approach. We omitted the Li example from Table 4.3 due to the fact that the Full simulation computation time was intractable.

The Pruned approach significantly reduces the computation time required for the exploration. Moreover, full simulation of the design space is often infeasible (due to prohibitive computation time). While in general, due to its heuristic nature, the pruned approach may not find all the points on the pareto curve, in practice it finds a large percentage of them, or approximates them well with close alternative designs. For instance, the coverage for the vocoder example shows that 83% of the designs on the pareto curve are successfully found by the Pruned exploration. While the Pruned approach does not find all the points on the pareto curve, the average difference between the points on the pareto and the corresponding closest points found by the exploration is 0.29% for cost,

Benchmark	Category	Pruned	Neighborhood	Full
compress	Time	2 days	2 weeks	1 month
	Coverage [%]	50%	65%	100%
	Avg. cost dist [%]	0.84%	0.59%	0%
	Avg. perf. dist [%]	0.77%	0.60%	0%
	Avg. energ. dist [%]	0.42%	0.28%	0%
vocoder	Time	24 min	29 min	50 min
	Coverage [%]	83%	100%	100%
	Avg. cost dist [%]	0.29%	0%	0%
	Avg. perf. dist [%]	2.96%	0%	0%
	Avg. energ. dist [%]	0.92%	0%	0%

Table 4.3.   Pareto coverage results for our Memory Architecture Exploration Approach.

2.96% for performance, and 0.92% for energy. In the compress example the computation time is reduced from 1 month for the Full simulation to 2 days, at the expense of less pareto coverage. However, while only 50% of the compress designs are exactly matched by the Pruned approach, for every pareto point missed, very close replacements points are generated, resulting in an average distance of 0.84%, 0.77%, and 0.42% in terms of cost, performance and power respectively, to the closest overall pareto point. Thus, our exploration strategy successfully finds most of the design points on the pareto curve without fully simulating the design space. Moreover, even if it misses some of the pareto points, it provides replacement architectures, which approximate well the pareto designs.

The Neighborhood exploration explores a wider design space than the Pruned approach, providing a better coverage of the pareto curve, at the expense of more computation time. For instance, for the Vocoder example, it finds 100% of the pareto points.

By performing combined exploration of the memory and connectivity architecture, we obtain a wide range of cost, performance and power tradeoffs. Clearly, these types of results are difficult to determine by analysis alone, and require a systematic exploration approach to allow the designer to best trade off the different goals of the system.

4.3.5   Related Work

There has been related work on connectivity issues in four main areas: (I) High-Level Synthesis, (II) System-on-Chip core-based systems, (III) Interface synthesis. and (IV) Layout and routing of connectivity wiring,

(I) In High-Level Synthesis, Narayan et al. [NG94] synthesize the bus structure and communication protocols to implement a set of virtual communication channels, trading off the width of the bus and the performance of the processes communicating over it. Daveau et al. [DIJ95] present a library based exploration approach, where they use a library of connectivity components, with different costs and performance. We complement these approaches by exploring the connectivity design space in terms of all the three design goals: cost, performance and power simultaneously.

Wuytack et. al. [WCdJ+96] present an approach to increase memory port utilization, by optimizing the memory mapping and code reordering. Our technique complements this work by exploring the connectivity architecture, employing connectivity components from an IP library. Catthoor et al. [CWG+98] address memory allocation, packing the data structures according to their size and bitwidth into memory modules from a library, to minimize the memory cost, and optimize port sharing. Forniciari et al. [FSS99] present a simulation based power estimation for the HW/SW communication on system-level busses, aimed at architectural exploration. We use the connectivity and memory power/area estimation models from [CWG+98] to drive our connectivity exploration.

(II) In the area of System-on-Chip architectures, Givargis et al. [GV98] present a connectivity exploration technique which employs different encoding techniques to improve the power behavior of the system. However, due to their platform-based approach, where they assume a pre-designed architecture platform which they tune for power, they do not consider the cost of the architecture as a metric. Maguerdichian et al. [MDK01] present an on-chip bus network design methodology, optimizing the allocation of the cores to busses to reduce the latency of the transfers across the busses. Lahiri et al. [LRLD00] present a methodology for the design of custom System-on-Chip communication architectures, which propose the use of dynamic reconfiguration of the communication characteristics, taking into account the needs of the application.

(III) Recent work on interface synthesis [COB95], [Gup95] present techniques to formally derive node clusters from interface timing diagrams. These techniques can be used to provide an abstraction of the connectivity and memory module timings in the form of Reservation Tables [HP90]. Our algorithm uses the Reservation Tables [GDN00, GHDN99] for performance estimation, taking into account the latency, pipelining, and resource conflicts in the connectivity and memory architecture.

(IV) At the physical level, the connectivity layout and wiring optimization and estimation has been addressed. Chen et al. [CZY+99] present a method to combine interconnect planning and floorplanning for deep sub-micron VLSI systems, where communication is increasingly important. Deng et al. [DM01] propose the use of a 2.5-D layout model, through a stack of single-layer monolithic ICs, to significantly reduce wire length. We use the area models presented in [CZY$^+$99] and [DM01] to drive our high-level connectivity exploration approach.

None of the previous approaches have addressed connectivity exploration in conjunction with memory modules architecture, considering simultaneously the cost, performance, and power of the system, using a library of connectivity components including standard busses (such as AMBA [ARM], mux-based connections, and off-chip busses). By pruning the non-interesting designs, we avoid simulating the complete design space, and allow the designer to efficiently target the system goals early in the design flow.

4.4   Discussion on Memory Architecture

This work has focused primarily on the software part of the Hardware/Software embedded system. While the Hardware/Software partitioning and Hardware synthesis have been extensively addressed by previous work [GVNG94, Gup95], we briefly discuss our Memory Exploration approach in the context of an overall Hardware/ Software architecture.

The input application can be implemented either in Hardware, in Software, or a combination of the two. We assume an initial step of Hardware/Software Partitioning decides which parts of the application are realized in Hardware and which in Software. The hardware parts are fed to a Hardware Synthesis step, generating the custom ASICs on the chip, while the software parts are fed to the compiler, producing code for the embedded processor.

There are three main issues when dealing with a system containing both software, and hardware ASICs, compared to a software-only system: (a) the synchronization and communication model between the software running on the CPU, and the ASIC, (b) the memory sharing model, and (c) the memory coherency mechanism.

(a) We assume two synchronization and communication models: (I) Shared memory, and (II) Message Passing. (I) In shared memory model, the synchronization and communication relies on a common memory space to store a set of semaphores and common data structures. The semaphores can be accessed through a set of atomic functions, and the hardware and software processes synchronize to insure sequentiality between dependent subtasks through a set of semaphore assignment and test functions. (II) In the message passing model, the ASIC and the programmable CPU communicate directly, through send-receive type protocols.

(b) We consider three memory models: (I) Shared memory model, (II) Partitioned memory model, and (III) Partially partitioned memory model. (I) In the shared memory model, we assume that both the CPU and the ASIC access a common memory address space. The synchronization needed to insure that the dependencies are satisfied can be implemented using either one of the two synchronization models presented above. (II) In the partitioned memory model, we assume that the memory spaces accessed by the ASIC and the CPU are disjoint. Each may have a local memory, but any communication between them relies on a form of message passing. (III) In the partially partitioned memory model, we assume that the ASIC and the CPU share a limited amount of memory space, and rely on local memories for the more data intensive computations.

(c) The memory coherency, or write coherency mechanism insures that the data is shared correctly across multiple memory modules, avoiding stale data. For instance, if the the CPU and the ASIC share a variable from the off-chip DRAM, but the CPU accesses it through the cache, while the ASIC reads directly from the DRAM, then if the CPU writes the variable in the cache, subsequent reads from the ASIC have to get the most recent value from the cache. Similarly, if the ASIC writes the variable in the DRAM, local copies in the cache may become stale, and have to be updated. In general, when a shared variable is written, all the copies in local memories have to be updated, or invalidated.

We assume two cache-coherence mechanisms to insure the coherence between different memory modules: (I) Write invalidate, and (II) Write update. In write invalidate, the writes which happen on the bus to the main memory are also found by the cache through snooping, and the corresponding lines are invalidated, setting an invalid bit. In write update, the cache line is updated when a write is recognized. If the cache is write-through, the notification of invalid data and the the actual data are sent at the same time. In write-back, the notification of invalid data is sent when the dirty bit is set, and the data is sent when the data is replaced in the cache, and the actual write occurs. A posible alternative to this approach is to asssume that writes to shared locations are write-through, and writes to non-shared locations are write-back (e.g., the DEC Firefly system [TSKS87]).

Figure 4.27 shows three example architectures, with varying memory and communication models. Figure 4.27 (a) shows a shared memory model, (b) shows a partially partitioned memory architecture, and (c) shows a partitioned memory architecture with message passing communication. The local memory in the architectures (b) and (c) may contain any combination of on-chip memory structures, such as caches, SRAMs, DMAs, or custom memory modules.

4.5 Summary and Status

In this chapter we presented an approach where by analyzing the access patterns in the application we gain valuable insight on the access and storage needs of the input application, and customize the memory architecture to better match these requirements, generating significant performance improvements for varied memory cost and power.

Traditionally, designers have attempted to alleviate the memory bottleneck by exploring different cache configurations, with limited use of more special purpose memory modules such as stream buffers [Jou90]. However, while re-

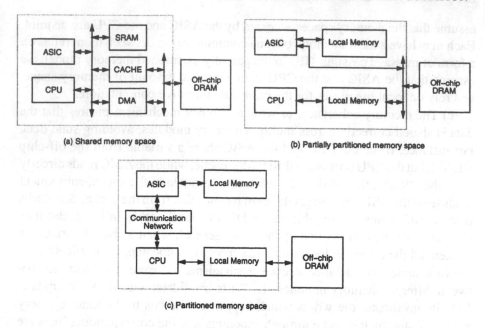

Figure 4.27.   Example memory model architectures.

alistic applications contain a large number of memory references to a diverse set of data structures, a significant percentage of all memory accesses in the application are generated from a few instructions, which often exhibit well-known, predictable access patterns. This presents a tremendous opportunity to customize the memory architecture to match the needs of the predominant access patterns in the application, and significantly improve the memory system behavior. We presented here such an approach called APEX that extracts, analyzes and clusters the most active access patterns in the application, and customizes the memory architecture to explore a wide range of cost, performance and power designs. We generate significant performance improvements for incremental costs, and explore a design space beyond the one traditionally considered, allowing the designer to efficiently target the system goals. By intelligently exploring the design space, we guide the search towards the memory architectures with the best cost/performance characteristics, and avoid the expensive full simulation of the design space.

Moreover, while the memory modules are important, often the connectivity between these modules have an equally significant impact on the system behavior. We presented our Connectivity Exploration approach (ConEx), which trades off the connectivity performance, power and cost, using connectivity modules from a library, and allowing the designer to choose the most promising connectivity architectures for the specific design goals.

We presented a set of experiments on large multimedia and scientific examples, where we explored a wide range of cost, performance and power tradeoffs, by customizing the memory architecture to fit the needs of the access patterns in the applications. Our exploration heuristic found the most promising cost/gain designs compared to the full simulation of the design space considering all the memory module allocations and access pattern cluster mappings, without the time penalty of investigating the full design space.

The memory and connectivity architecture exploration approach presented in this chapter has been implemented. The experimental results have been obtained using our memory simulator based the SIMPRESS [MGDN01] memory model, and SHADE [CK93].

Chapter 5

MEMORY-AWARE COMPILATION

5.1 Motivation

In recent SOC and processor architectures, memory is identified as a key performance and power bottleneck [Prz97]. With advances in memory technology, new memory modules (e.g., SDRAM, DDRAM, RAMBUS, etc.) that exhibit efficient access modes (such as page mode, burst mode, pipelined access [Pri96], [Prz97]) appear on the market with increasing frequency. However, without compiler support, such novel features cannot be efficiently used in a programmable system.

Furthermore, in the context of Design Space Exploration (DSE), the system designer would like to evaluate different combinations of memory modules and processor cores from an IP library, with the goal of optimizing the system for varying goals such as cost, performance, power, etc.. In order to take full advantage of the features available in each such processor-memory architecture configuration, the compiler needs to be aware of the characteristics of each memory library component.

Whereas optimizing compilers have traditionally been designed to exploit special architectural features of the processor (e.g., detailed pipelining information), there is a lack of work addressing memory-library-aware compilation tools that explicitly model and exploit the high-performance features of such diverse memory modules. Indeed, particularly for memory modules, a more accurate timing model for the different memory access modes allows for a better match between the compiler and the memory sub-system, leading to better performance.

Moreover, with the widening gap between processor and memory latencies, hiding the latency of the memory operations becomes increasingly important. In particular, in the presence of caches, cache misses are the most time-consuming operations (orders of magnitude longer than cache hits), being responsible for considerable performance penalties.

In this chapter we present an approach that allows the compiler to exploit detailed timing information about the memory modules, by intelligently managing the special access modes (such as page-mode, burst-mode, etc.), and cache miss traffic, and better hide the latency of the lengthy off-chip memory transfers. In Section 5.2 we present our Timing Extraction algorithm, and its use in the optimization of the memory transfers exhibiting efficient access modes. In Section 5.3 we present the utilization of accurate timing information in the presence of caches. Section 5.4 concludes with a short summary.

5.2   Memory Timing Extraction for Efficient Access Modes

Traditionally, efficient access modes exhibited by new memory modules (such as SDRAM, DDRAM, RAMBUS, etc.) were transparent to the processor, and were exploited implicitly by the memory controller (e.g., whenever a memory access referenced an element already in the DRAM's row buffer, it avoided the row-decode step, fetching it directly from the row buffer). However, the memory controller only has access to local information, and is unable to perform more global optimizations (such as global code reordering to better exploit special memory access modes). By providing the compiler with a more accurate timing model for the specific memory access modes, it can perform global optimizations that effectively hide the latency of the memory operations, and thereby generate better performance.

In this section we describe an approach that exposes the detailed timing information about the memory modules to the compiler, providing an opportunity to perform global optimizations. The key idea is that we combine the timing model of the memory modules (e.g., efficient memory access modes) with the processor pipeline timings to generate accurate operation timings. We then use these exact operation timings to better schedule the application, and hide the latency of the memory operations.

In Section 5.2.1 we show a simple motivating example to illustrate the capabilities of our technique, and in Section 5.2.2 we present the flow of our approach. In Section 5.2.3 we describe our timing generation algorithm. In Section 5.2.4 we present a set of experiments that demonstrate the use of accurate timing for the TIC6201 architecture with a synchronous DRAM module, and present the performance improvement obtained by our memory-aware compiler. In Section 5.2.5 we present previous work addressing memory and special access mode related optimizations.

5.2.1 Motivating Example

Burst mode access is a typical efficient access mode for contemporary DRAMs (e.g., SDRAM) that is not fully exploited by traditional compilers. We use a simple example to motivate the performance improvement made possible by compiler exploitation of such access modes through a more accurate timing model.

The sample memory library module we use here is the IBM0316409C [IBM] Synchronous DRAM. This memory contains 2 banks, organized as arrays of 2048 rows x 1024 columns, and supports normal, page mode, and burst mode accesses. A normal read access starts by a row decode (activate) stage, where the entire selected row is copied into the row buffer. During column decode, the column address is used to select a particular element from the row buffer, and output it. The normal read operation ends with a precharge (or deactivate) stage, wherein the data lines are restored to their original values.

For page mode reads, if the next access is to the same row, the row decode stage can be omitted, and the element can be fetched directly from the row buffer, leading to a significant performance gain. Before accessing another row, the current row needs to be precharged.

During a burst mode read, starting from an initial address input, a number of words equal to the burst length are clocked out on consecutive cycles without having to send the addresses at each cycle.

Another architectural feature which leads to higher bandwidth in this DRAM is the presence of two banks. While one bank is bursting out data, the other can perform a row decode or precharge. Thus, by alternating between the two banks, the row decode and precharge times can be hidden. Traditionally, the architecture would rely on the memory controller to exploit the page/burst access modes, while the compiler would not use the detailed timing model. In our approach, we incorporate accurate timing information into the compiler, which allows the compiler to exploit more globally such parallelism, and better hide the latencies of the memory operations.

We use the simple example in Figure 5.1 (a) to demonstrate the performance of the system in three cases: (I) without efficient access modes, (II) optimized for burst mode accesses, but without without the compiler employing an 1accurate timing model, and (III) optimized for burst mode accesses with the compiler exploiting the accurate timing model.

The primitive access mode operations [PDN99] for a Synchronous DRAM are shown in Figure 5.1 (b): the un-shaded node represents the row decode operation (taking 2 cycles), the solid node represents the column decode (taking 1 cycle), and the shaded node represents the precharge operation (taking 2 cycles).

Figure 5.1 (c) shows the schedule for the unoptimized version, where all reads are normal memory accesses (composed of a row decode, column decode, and precharge). The dynamic cycle count here is 9 x (5 x 4) = 180 cycles.

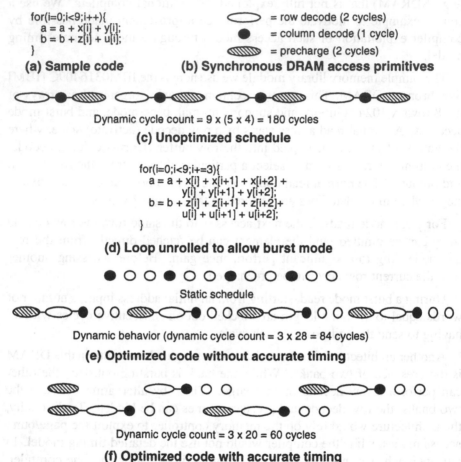

```
for(i=0;i<9;i++){
    a = a + x[i] + y[i];
    b = b + z[i] + u[i];
}
```
(a) Sample code

◯ = row decode (2 cycles)
● = column decode (1 cycle)
▧ = precharge (2 cycles)

(b) Synchronous DRAM access primitives

Dynamic cycle count = 9 x (5 x 4) = 180 cycles
(c) Unoptimized schedule

```
for(i=0;i<9;i+=3){
    a = a + x[i] + x[i+1] + x[i+2] +
        y[i] + y[i+1] + y[i+2];
    b = b + z[i] + z[i+1] + z[i+2]+
        u[i] + u[i+1] + u[i+2];
}
```
(d) Loop unrolled to allow burst mode

Static schedule

Dynamic behavior (dynamic cycle count = 3 x 28 = 84 cycles)
(e) Optimized code without accurate timing

Dynamic cycle count = 3 x 20 = 60 cycles
(f) Optimized code with accurate timing

Figure 5.1.   Motivating example

In order to increase the data locality and allow burst mode access to read consecutive data locations, an optimizing compiler would unroll the loop 3 times. Figure 5.1 (d) shows the unrolled code, and Figure 5.1 (e) shows the static schedule, and the dynamic (run-time) schedule of the code [1], for a schedule with no accurate timing. Traditionally, the memory controller would handle all the special access modes implicitly, and the compiler would schedule the code optimistically, assuming that each memory access takes 1 cycle (the length of a

---

[1] In Figure 5.1 (c) the static schedule and the run-time behavior were the same. Here, due to the stalls inserted by the memory controller, they are different

page mode access). During a memory access which takes longer than expected, the memory controller has to freeze the pipeline, to avoid data hazards. Thus, even though the static schedule seems faster, the dynamic cycle-count in this case is 3 x 28 = 84 cycles.

Figure 5.1 (f) shows the effect of scheduling using accurate memory timing on code that has already been optimized for burst mode. Since the memory controller does not need to insert stalls anymore, the dynamic schedule is the same as the static one (shown in Figure 5.1 (f)). Since accurate timing is available, the scheduler can hide the latency of the precharge and row decode stages, by for instance precharging at the same time the two banks, or executing row decode while the other bank bursts out data. The dynamic cycle count here is 3 x 20 = 60 cycles, resulting in a 40% improvement over the best schedule a traditional optimizing compiler would generate.

Thus, by providing the compiler with more detailed information, the efficient memory access modes can be better exploited. The more accurate timing model creates a significant performance improvement, in addition to the page/burst mode optimizations.

5.2.2 Our Approach

Figure 5.2 outlines our IP library based Design Space Exploration (DSE) scenario. The processor-memory architecture is captured using components from the processor and memory IP libraries in the EXPRESSION Architecture Description Language (ADL) [HGG+99]. This architectural description is then used to target the compilation process to the specific processor-memory architecture chosen by the designer. In order to have a good match between the compiler and the architecture, detailed information about resources, memory features, and timing has to be provided.

As shown in Figure 5.2, the timing information for the compiler is generated using two parts: (1) RTGEN, which extracts the resources of operations that flow through the processor pipeline, using reservation tables, and (2) TIMGEN, which extracts the timing of memory operations flowing through the memory pipeline. For any operations executing on the processor, our compiler combines these two timings to effectively exploit the pipeline structures both within the processor, as well as within the memory subsystem.

Since contemporary high-end embedded processors contain deep pipelines and wide parallelism, hazards in the pipeline can create performance degradation, or even incorrect behavior, if undetected. There are three types of hazards in the pipeline: resource, data, and control hazards [HP90]. RTGEN [GHDN99], shown in Figure 5.2, is a technique that automatically generates reservation tables containing detailed resource information for execution of operations on a pipelined processor.

Data hazards occur when the pipeline changes the order of read/write accesses to operands, breaking data dependencies [HP90]. In this chapter we address the data hazards problem (the TIMGEN algorithm in Figure 5.2), by marrying the timing information from memory IP modules with the timing information of the processor pipeline architecture, and generate the operation timings needed by the compiler to avoid such data hazards.

Our RTGEN algorithm thus addresses the resource hazards problem by automatically generating the detailed Reservation Tables needed to detect structural hazards in the pipeline.

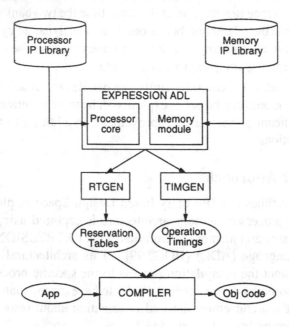

Figure 5.2.   The Flow in our approach

Since memory is a major bottleneck in the performance of such embedded systems, we then show how this accurate timing model can be used by the compiler to better exploit memory features such as efficient access modes (e.g., page-mode and burst-mode), to obtain significant performance improvements.

Furthermore, we provide the system designer with the ability to evaluate and explore the use of different memory modules from the IP library, with the memory-aware compiler fully exploiting the special efficient access modes of each memory component.

The next section outlines the TIMGEN algorithm.

5.2.3   TIMGEN: Timing extraction algorithm

Operation timings have long been used to detect data hazards within pipelined processors [HP90]. However, compilers have traditionally used fixed timings for scheduling operations. In the absence of dynamic (run-time) data hazard detection capabilities in the hardware (such as in VLIW processors), the compiler has to use conservative timings, to avoid such hazards, and ensure correct behavior. In the presence of dynamic control capabilities the compiler may alternatively use optimistic timings, and schedule speculatively, knowing that the hardware will account for the potential delay.

For memory-related operations, the memory controller provides some dynamic control capabilities, which can for instance stall the pipeline if a load takes a long time. In this case a traditional compiler would use fixed timings to schedule either conservatively or optimistically the memory operations. In the conservative approach, the compiler assumes that the memory contains no efficient access modes, and all accesses require the longest delay. Even though this results in correct code, it wastes a lot of performance. Alternatively, the optimistic schedule is the best the compiler could do in the absence of accurate timing: the compiler assumes that all the memory accesses have the length of the fastest access (e.g., page mode read), and relies on the memory controller to account for the longer delays (e.g., by freezing the pipeline). Clearly, this creates an improvement over the conservative schedule. However, even this optimistic approach can be significantly improved by integrating a more accurate timing model for memory operations. Indeed, a memory-aware compiler can better exploit the efficient memory access modes of a memory IP by hiding the latency of the memory operations. This creates an opportunity for considerable further performance improvements.

As shown in Figure 5.2, our timing generation algorithm, called TIMGEN, starts from an EXPRESSION ADL [HGG+99] containing a structural description of the processor pipeline architecture, and an abstraction of the memory module access mode timings, and generates the operation timings, consisting of the moments in time (relative to the issue cycle of the operation) when each operand in that operation is read (for sources) or written (for destinations).

To illustrate the TIMGEN algorithm, we use the example architecture in Figure 5.3, based on the TI TMS320C6201 VLIW DSP, with an SDRAM block attached to the External Memory Interface (EMIF). For the sake of illustration, Figure 5.3 presents the pipeline elements and the memory timings only for load operations executed on the D1 functional unit.

The TIC6201 processor contains 8 functional units, an 11-stage pipeline architecture, and no data cache. The pipeline contains: 4 fetch stages (PG, PS, PW, PR), 2 decode stages (DP, DC), followed by the 8 pipelined functional units (L1, M1, S1, D1, L2, M2, S2 and D2). In Figure 5.3 we present the D1 load/store functional unit pipeline with 5 stages: D1_E1, D1_E2, EMIF, MemCtrl_E1 and MemCtrl_E2. The DRAM1 represents the SDRAM memory module attached to the EMIF, and RFA represents one of the two TIC6201 distributed register files.

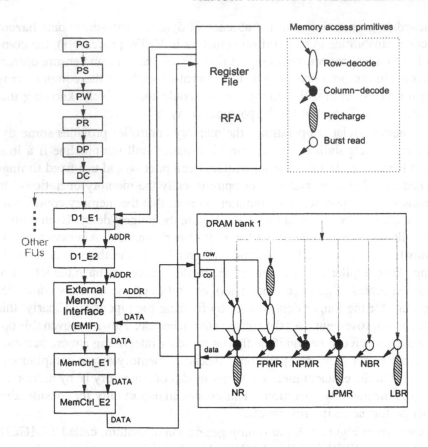

Figure 5.3. Example architecture, based on TI TMS320C6201

A typical memory operation flows through the processor pipeline until it reaches the D1_E1 stage. The D1_E1 stage computes the load/store address, D1_E2 transfers the address across the CPU boundary, to the EMIF, which sends the read/write request to the memory module. Since the SDRAM access may incur additional delay depending on the memory access mode, the EMIF will wait until the SDRAM access is finished. MemCtrl_E1 then transfers the data received from EMIF back into the CPU, and MemCtrl_E2 writes it to the destination register.

The graphs inside the DRAM1 of Figure 5.3 represent the SDRAM's memory timings as an abstraction of the memory access modes, through combinations of a set of primary operation templates corresponding to the row-decode, column-decode and precharge (similar to Figure 5.1 (b)). Normal Read (NR) represents a normal memory read, containing a row-decode, column-decode and precharge. First Page Mode (FPMR) represents the first read in a sequence of page-mode reads, which contains a row precharge, a row-decode and a column decode. The next page mode operations in the sequence, shown as Next Page Mode (NPMR), contain only a column-decode. Similarly, Next Burst (NBR) represents burst-mode reads which contain only the bursting out of the data from the row-buffer. The Last Page Mode (LPMR) and Last Burst Mode (LBR) Read are provided to allow precharging the bank at the moment when the page-mode or burst sequence ends (instead of waiting until the next sequence starts), thus allowing for more optimization opportunities.

The TIMGEN algorithm is outlined in Figure 5.4. The basic idea behind the TIMGEN algorithm is that during execution, an operation proceeds through a pipeline path, and accesses data through some data transfer paths. For memory operations, the data transfer paths can access complex memory modules (e.g., SDRAM), and the operation may be delayed based on the memory access mode (e.g., normal/page/burst mode). Therefore, we can collect detailed timing information regarding each operand for that operation, by tracing the progress of the operation through the pipeline and storage elements.

To demonstrate how the TIMGEN algorithm works, we use the load operation *LDW .D1 RFA0[RFA1], RFA2*, executed on the architecture shown in Figure 5.3. Assuming this load is a first read in a sequence of page-mode reads, we want to find in what cycle the base argument (RFA0), offset argument (RFA1), and the implicit memory block argument (the DRAM bank) are read, and in what cycle the destination register argument (RFA2) is written.

Detailed timing is computed by traversing the pipeline paths, data transfer paths, and complex memory modules, and collecting the cycle where each argument is read/written. In this instance, the *LDW .D1 RFA0[RFA1], RFA2* operation starts by traversing the 4 fetch pipeline stages (PG, PS, PW, PR), and 2 decode stages (DP, DC). Then, the operation is directed to the stage 1 of the D1 functional unit (D1_E1) [2]. Here, the two source operands, representing the base register and the offset register, are read from the RFA register file (RFA0 and RFA1), and the address is computed by left-shifting the offset by 2, and adding it to the base. Thus, the base and offset are read during the stage 7 of the pipeline.

During stage D1_E2, the address is propagated across the CPU boundaries, to the External Memory Interface (EMIF). Here, the EMIF sends the address to the SDRAM block, which starts reading the data. Depending on the type of memory access (normal/page/burst) and on the context information (e.g., whether the row is already in the row buffer), the SDRAM block may require a different amount of time. Assuming that this load operation is the first in a sequence of page mode reads, TIMGEN chooses the First Page Mode (FPM)

---

[2]For the sake of clarity, the other functional units have not been represented

Algorithm: TIMGEN
Input: Processor pipeline and memory access mode timings description,
a memory operation, and the access_mode used (normal/page/burst)
Output: Timings for all operation arguments
Begin TIMGEN
  For curr_pipe_stage traversing the pipeline stages used by the input operation
  (in order of execution)
    For all the data_transfer_paths accesed in the curr_pipeline_stage
      Find the operation_argument corresponding to this data_transfer_path
      Find the storage_element accessed by this data_transfer_path
      If storage_element is a complex storage module (e.g., an SDRAM module)
        Choose the access_pattern in this storage_element, according to the access_mode
        Find the cycle when the operation_argument is transfered in this access_pattern
        Add (cycle + curr_pipe_stage) as the timing for the operation_argument
      Else If storage_element is a simple storage module (e.g., register file)
        Add the curr_pipe_stage as the timing for the operation_argument
      Update the curr_pipe_stage
End TIMGEN

Figure 5.4. The TIMGEN timing generation algorithm.

template from the SDRAM block. Here we need to first precharge the bank, then perform a row-decode followed by a column decode. During the row-decode, the row part of the address is fed to the row decoder, and the row is copied into the row buffer. During column-decode, the column part of the address is fed to the column decoder, and the corresponding data is fetched from the row buffer. Since the prefetch and row-decode require 2 cycles each, and the column-decode requires 1 cycle, the data will be available for the EMIF 5 cycles after the EMIF sent the request to the SDRAM block. Thus the data is read 14 cycles after the operation was issued.

During the MemCtrl_E1 stage, the data is propagated back across the CPU boundary. The MemCtrl_E2 stage writes the data to the register file destination (RFA2), 16 cycles after the operation issue.

In this manner, the TIMGEN algorithm (Figure 5.4) computes the detailed timings of all the arguments for that given operation: the base argument (RFA0) and the offset argument (RFA1) are read during stage 7 of the pipeline, the memory block (the SDRAM array) is read during stage 14, and the destination argument (RFA2) is written during stage 16. A detailed description of the algorithm can be found in [GDN99].

The worst case complexity of TIMGEN is $O(x*y)$, where x is the maximum number of pipeline stages in the architecture, and y the maximum number of data transfers in one pipeline stage. Most contemporary pipelined architectures have between 5 and 15 stages in the pipeline, and 1 to 3 data transfer paths per pipeline stage, leading to reasonable computation time when applied to different pipeline architectures and memory modules with varying access modes. Since during compilation, the timings for all the operations in the application are required, we can compute before-hand these timings, and store them in a database.

5.2.4 Experiments

We now present a set of experiments demonstrating the performance gains obtained by using accurate timing in the compiler. We first optimize a set of benchmarks to better utilize the efficient memory access modes (e.g., through memory mapping, code reordering or loop unrolling [3] [PDN98]), and then we use the accurate timing model to further improve the performance by hiding the latencies of the memory operations. To separate out the benefit of the better timing model from the gain obtained by a memory controller exploiting the access mode optimizations and the access modes themselves, we present and compare two sets of results: (1) the performance gains obtained by scheduling with accurate timing in the presence of a code already optimized for memory accesses, and (2) the performance of the same memory-access-optimized code (using less accurate timing) scheduled optimistically, assuming the shortest access time available (page-mode access), and relying on the memory controller to account for longer delays. This optimistic scheduling is the best alternative available to the compiler, short of an accurate timing model. We also compare the above approaches to the performance of the system in the absence of efficient memory access modes.

5.2.4.1 Experimental Setup

In our experiments, we use an architecture based on the Texas Instruments TIC6201 VLIW DSP, with one IBM0316409C synchronous DRAM [IBM] block exhibiting page-mode and burst-mode access, and 2 banks. The TIC6201 is an integer-point 8-way VLIW processor, with no data cache, as explained earlier. The External Memory Interface (EMIF) [Tex] allows the processor to program information regarding the memory modules attached, and control them through a set of control registers. We assume the SDRAM has the capability to precharge a specific memory blank (using the DEAC command), or both memory banks at the same time (using the DCAB command), and to perform a row decode while the other bank bursts out data (using the ACTV command).

The applications have been compiled using our EXPRESS [HDN00] retargetable optimizing ILP compiler. Our scheduler reads the accurate timing model generated by the TIMGEN algorithm, and uses Trailblazing Percolation Scheduling (TiPS) [NN93] to better target the specific architecture. TiPS is a powerful Instruction Level Parallelism (ILP) extraction technique, which can

---

[3] Loop unrolling trades-off performance against code size and register pressure [PDN98] (we assumed 32 registers available).

fully exploit the accurate operation timings, by highly optimizing the schedule. The cycle counts have been computed using our structural cycle-accurate simulator SIMPRESS [KSH+99].

5.2.4.2   Results

The first column in Table 5.1 shows the benchmarks we executed (from multimedia and DSP domains). The second column shows the dynamic cycle counts when executing the benchmarks on the architecture using only normal memory accesses (no efficient memory access modes, such as page/burst mode).

The third column represents the dynamic cycle counts optimized for efficient access modes, but with no accurate timing model. For fairness of the comparison we compile using the optimistic timing model, and rely on the memory controller to stall the pipeline when the memory accesses take longer then expected. Notice that by using these efficient access modes, and optimizing the code to better exploit them (e.g., by loop unrolling), we obtain a very high performance gain over the baseline unoptimized case.

The fourth column represents the dynamic cycle counts using both efficient access modes, and using an accurate timing model to better hide the latency of the memory operations. By comparing these figures to the previous column, we separate out the performance gains obtained by using an accurate timing model, in addition to the gains due to the efficient access mode optimizations. The fifth column shows the percentage performance improvement of the results for the code optimized with accurate timing model (fourth column), compared to the results optimized without accurate timing model (third column).

The performance gains from exploiting detailed memory timing vary from 6% (in GSR, where there are few optimization opportunities), to 47.9% (in SOR, containing accesses which can be distributed to different banks and memory pages), and an average of 23.9% over a schedule that exploits the efficient access modes without detailed timing.

While the core of our TIMGEN optimization (that uses accurate timing information in the compiler to improve the performance) does not generate code size increase, the code transformations performed by the traditional optimizing compiler (such as loop unrolling [PDN98]) may generate increased code size. Table 5.2 presents the code size increase generated by the traditional loop transformations. The first column shows the benchmarks. The second column shows the number of assembly lines for the unoptimized base-line code, and the third column shows the number of assembly lines after the optimizing compiler performs code transformations (such as loop unrolling and code reordering). The last column shows the percentage code size increase over the unoptimized code. The code size increase varies between 0%, for SOR, where

Benchmark	Unoptimized (normal accesses)	Optimized (page/burst mode)		
		Optimized code w/o accurate timing	Optimized code w/ accurate timing	% perf gain
SOR	192286	104350	70510	47.9
GSR	176034	111973	105346	6
beam	439087	41734	30854	35.2
idct	27163	8603	6939	23.9
mm	12909	3336	2568	29.9
dhrc	13957	6277	5637	10
madd	4005	987	727	35.7
dequant	6105	5105	4605	10
leaf_plus	2905	2405	1955	23
lowpass	7433	4169	3529	18
			Average	**23.9**

Table 5.1. Dynamic cycle counts for the TIC6201 processor with an SDRAM block exhibiting 2 banks, page and burst accesses

Benchmark	Unoptimized code size	Optimized code size	Code size increase [%]
SOR	142	142	0
GSR	118	158	33.89
beam	137	149	8.75
idct	157	352	124.20
mm	83	193	132.53
dhrc	130	130	0
madd	61	113	85.24
dequant	113	113	0
leafplus	82	82	0
lowpass	136	166	22.05
		Average	**40.6**

Table 5.2. Number of assembly lines for the first phase memory access optimizations

enough page-mode and burst-mode accesses were already present in the code without program transformations, and 132%, for mm, where aggresive loop unrolling has been performed.

While these code-expanding transformations improve the locality of the accesses to the DRAM page, they are not always needed. Even in the absence of such transformations, directly using the efficient access modes such as page- and burst-mode accesses, results in significant performance improvements (e.g.,

SOR, dhrc, dequant, leafplus). Moreover, for area-constrained embedded systems, the designer can trade-off code size against performance, by limiting the unroll factor. Alternatively, when the access patterns are uniform across the application, memory allocation optimizations that increase the page-locality (such as interleaving the simultaneously accessed arrays [PDN99]), can be used to replace the loop transformations, thus avoiding the code size increase.

The large performance gains obtained by using the more accurate operation timings are due to the better opportunities to hide the latency of the lengthy memory operations, by for instance performing row-decode, or precharge on one bank, while the other bank is bursting out data, or using normal compute operations to hide the latency of the memory operations. This effect is particularly large in the benchmarks where there are many operations available to hide in parallel with the memory operations (e.g., SOR, beam, IDCT, etc.). In cases when the memory access patterns do not allow efficient exploitation of the SDRAM banks (e.g., GSR), the gain is smaller. However, we are able to exploit accurate timing information for many multimedia and DSP applications, since they usually contain multiple arrays, often with two or more dimensions, which can be spread out over multiple memory banks.

5.2.5   Related Work

Related work on memory optimizations has been addressed previously both in the custom hardware synthesis and in the embedded processor domains. In the context of custom hardware synthesis, several approaches have been used to model and exploit memory access modes. Ly et. al. [LKMM95] use behavioral templates to model complex operations (such as memory reads and writes) in a CDFG, by enclosing multiple CDFG nodes and fixing their relative schedules (e.g., data is asserted one cycle after address for a memory write operation). However, since different memory operations are treated independently, the scheduler has to assume a worst-case delay for memory operations – missing the opportunity to exploit efficient memory access modes such as page and burst mode.

Panda et. al. [PDN98] outline a pre-synthesis approach to exploit efficient memory access modes, by massaging the input application (e.g., loop unrolling, code reordering) to better match the behavior to a DRAM memory architecture exhibiting page-mode accesses. Khare et. al. [KPDN98] extend this work to Synchronous and RAMBUS DRAMs, using burst-mode accesses, and exploiting memory bank interleaving.

Wuytack et. al. [WCdJ$^+$96] present an approach to increase memory port utilization, by optimizing the memory mapping and code reordering. Our technique complements this work by exploiting the efficient memory access modes to further increase the memory bandwidth.

While the above techniques work well for custom hardware synthesis, they cannot be applied directly to embedded processors containing deep pipelines and wide parallelism: a compiler must combine the detailed memory timing with the processor pipeline timing to create the timing for full-fledged operations in the processor-memory system.

Processors traditionally rely on a memory controller to synchronize and utilize specific access modes of memory modules (e.g., freeze the pipeline when a long delay from a memory read is encountered). However, the memory controller only has a local view of the (already scheduled) code being executed. In the absence of an accurate timing model, the best the compiler can do is to schedule optimistically, assuming the fastest access time (e.g., page mode), and rely on the memory controller to account for longer delays, often resulting in performance penalty. This optimistic approach can be significantly improved by integrating an accurate timing model into the compiler. In our approach, we provide a detailed memory timing model to the compiler so that it can better utilize efficient access modes through global code analysis and optimizations, and help the memory subsystem produce even better performance. We use these accurate operation timings in our retargetable compiler to better hide the latency of the memory operations, and obtain further performance improvements.

Moreover, in the absence of dynamic data hazard detection (e.g., in VLIW processors), these operation timings are *required* to insure correct behavior: the compiler uses them to insert NOPs in the schedule to avoid data hazards. In the absence of a detailed timing model, the compiler is forced to use a pessimistic schedule, thus degrading overall performance.

In the embedded and general purpose processor domain, a new trend of instruction set modifications has emerged, targeting explicit control of the memory hierarchy, through for instance prefetch, cache freeze, and evict-block operations (e.g., TriMedia 1100, StrongArm 1500, IDT R4650, Intel IA 64, Sun UltraSPARC III, etc. [hot]). Even though such operations can improve memory traffic behavior, they are orthogonal to the specific library modules used. By exploiting detailed timing of efficient memory access modes, we can provide further performance improvement opportunities.

In the domain of programmable SOC architectural exploration, recently several efforts have proposed the use of Architecture Description Languages (ADLs) to drive generation of the software toolchain (compilers, simulators, etc.) ([HD97], [Fre93], [Gyl94], [HGG+99], [LM98]). However, most of these approaches have focused primarily on the processor and employ a generic model of the memory subsystem. For instance, in the Trimaran compiler [Tri97], the scheduler uses operation timings specified on a per-operation basis in the MDes ADL to better schedule the applications. However they use fixed operation timings, and do not exploit efficient memory access modes. Our approach uses

EXPRESSION [HGG⁺99], a memory-aware ADL that explicitly provides a detailed memory timing to the compiler and simulator.

Moreover, in the context of Design Space Exploration (DSE), when different processor cores and memory IP modules are mixed and matched to optimize the system for different goals, describing the timings on a per-operation basis requires re-specification whenever the pipeline architecture or the memory module is changed. Since changes in the pipeline and memory architecture impact the timings of all operations (not only memory related), updating the specification during every DSE iteration is a very cumbersome and error prone task. Instead, in our approach we extract these operation timings from the processor core pipeline timing and the storage elements timings, allowing for faster DSE iterations. No previous compiler can exploit detailed timing information of efficient memory access modes offered by modern DRAM libraries (e.g., SDRAM, RAMBUS).

Recent work on interface synthesis [COB95], [Gup95] present techniques to formally derive node clusters from interface timing diagrams. These techniques can be applied to provide an abstraction of the memory module timings required by our algorithm.

Our technique generates accurate operation timing information by marrying the pipeline timing information from the processor core module, with timing information from the memory library modules, and uses it in a parallelizing compiler to better target the processor-memory system. We exploit efficient memory access modes such as page-mode and burst-mode, to increase the memory bandwidth, and we use the accurate timing information to better hide the latency of the memory operations, and create significant performance improvements. We support fast DSE iterations, by allowing the designer to plug in a memory module from an IP library, and generate the operation timings, in a compiler-usable form.

5.3   Memory Miss Traffic Management

Traditional optimizing compilers focused mainly on cache hit accesses (e.g., cache hit ratio optimizations), but did not actively manage the cache miss traffic. They optimistically treated all the memory accesses as cache hits, relying on the memory controller to hide the latency of the longer cache miss accesses. However, not all cache misses can be avoided (e.g., compulsory misses). Moreover, since the memory controller has only access to a local view of the program, it cannot efficiently hide the latency of these memory accesses. This optimistic approach can be significantly improved by integrating an accurate timing model into the compiler, and hiding the latency of the longer miss accesses. By providing the compiler with cache hit/miss traffic information and an accurate timing model for the hit and miss accesses, it is possible to perform global optimizations and obtain significant performance improvements. We have not seen any previous work addressing memory-aware compilation tools that focus on the cache miss traffic, and specifically target the memory bandwidth, by using cache behavior analysis and accurate timing to aggressively overlap cache miss transfers.

The main contribution is the idea of explicitly managing the cache misses from the code to better utilize the memory bandwidth, and better overlap them with the cache hits and CPU operations. Since the misses are the most time-consuming operations, and represent the bottleneck in memory-intensive applications, by focusing on them first, we can reduce the memory stalls and improve performance.

In Section 5.3.1 we show a simple motivating example to illustrate the capabilities of our technique, and in Section 5.3.2 we describe our miss traffic management algorithm. In Section 5.3.3 we present a set of experiments that demonstrate the explicit management of main memory transfers for the TIC6211 architecture with a cache-based memory architecture, and present the performance improvements obtained by our memory-aware compiler. In Section 5.3.4 we present previous work addressing cache and memory optimizations.

5.3.1   Illustrative example

We use the simple illustrative example in Figure 5.5 (a) to demonstrate the performance improvement of our technique. In this example we assume a cache hit latency of 2 cycles, cache miss latency of 20 cycles, and non-blocking writes which take 2 cycles. We use a 16KB direct mapped cache with lines of 4 words, each word 4 bytes. For clarity of the explanation we assume no parallelism or pipelining available between cache misses[4], but cache hits can be serviced at the same time as cache misses.

We present the example code in three cases: (I) the traditional approach, where the compiler uses optimistic timing to schedule the memory operations, assuming they are all hits, and relies on the memory controller to account for misses, (II) the first phase of the miss traffic optimization, with cache miss traffic analysis and accurate miss timing, and (III) the second phase of the miss traffic optimization, with cache analysis and accurate timing, as well as aggressively optimized miss traffic schedule.

The primitive operation nodes used in Figure 5.5 represent the cache miss, cache hit, addition, and memory write operations. The shaded nodes grouping several such primitive nodes represent loop iterations.

I. In the absence of accurate timing information, the best the compiler can do is to treat all the memory accesses as hits, and schedule them optimistically, relying on the memory controller to account for longer delays. Figure 5.5

---

[4]Our model however allows us to describe and exploit any level of pipelining and parallelism in the memory subsystem [GDN00], [GHDN99]

(b) presents the static schedule and the dynamic behavior for the traditional optimistic scheduling approach. In the example from Figure 5.5 (a) every fourth access to the arrays a and b results in a cache miss. As a result, in every fourth iteration the memory controller has to insert stalls to account for the latency of the cache misses, generating a total dynamic cycle count of 256 cycles.

II. By analyzing the cache behavior, and providing accurate timing information for all the accesses, the compiler can better hide the latency of the cache misses, and improve overall performance. Since we know that every fourth access to arrays a and b results in a miss, we unroll the loop 4 times (shown in Figure 5.5 (c)), and isolate the cache misses, to allow the compiler to attach accurate timing information to the memory accesses. The first 2 accesses in the unrolled loop body (a[i] and b[i]) result in cache misses, while all the other accesses result in hits. The compiler can attach accurate timing to the hit and miss accesses, to schedule the operations and hide some of the latencies associated with the memory operations. Figure 5.5 (d) shows the dynamic behavior [5] for the code in Figure 5.5 (c) scheduled using accurate timing and cache behavior information. The dynamic cycle count in this case is 220 cycles, resulting in a 16% improvement over the traditional approach.

III. After performing cache analysis and attaching accurate timing information to the memory accesses, we aggressively schedule the miss traffic to create more opportunity for memory access overlap. By recognizing that often (especially after cache locality optimizations have been applied), accesses to the same cache line are close together in the code, the performance can be further improved. Since the first access to a cache line generates a miss and the subsequent accesses to that line have to wait for that miss to complete, the compiler is limited in its capability to efficiently schedule these operations (we will introduce in Section 5.3.2 the notion of "cache dependency", which captures such information). However, by overlapping the cache miss to one cache line with cache hits to a different cache line, it is possible to increase the potential parallelism between memory operations, generating further performance improvements. For instance, the cache miss a[i] from Figure 5.5 (c) accesses the same cache line as the hits in a[i+1], a[i+2], and a[i+3], and requires them to wait for the completion of the transfer from the main memory. By shifting the array accesses a[i] and b[i] from iteration i to the previous iteration, i-1, we allow for more parallelism opportunities. Figure 5.5 (e) shows the unrolled and shifted code, which further optimizes the cache miss traffic by overlapping the cache misses to one cache line with the cache hits from another cache line.

Figure 5.5 (f) shows the dynamic behavior for the unrolled and shifted code, optimized using accurate timing and cache behavior information to aggressively

---

[5]The static schedule and the dynamic execution in this case are the same

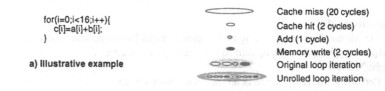

```
for(i=0;i<16;i++){
    c[i]=a[i]+b[i];
}
```

a) Illustrative example

⬭ Cache miss (20 cycles)
○ Cache hit (2 cycles)
∘ Add (1 cycle)
● Memory write (2 cycles)
○○∘● Original loop iteration
Unrolled loop iteration

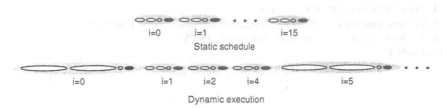

i=0   i=1   i=15

Static schedule

i=0   i=1   i=2   i=4   i=5

Dynamic execution

b) Static schedule and dynamic execution for the original example
(dynamic cycle count = (43+7*3)*(16/4) = 256 cycles)

```
for(i=0;i<16;i+=4){
    c[i]=a[i]+b[i];
    c[i+1]=a[i+1]+b[i+1];
    c[i+2]=a[i+2]+b[i+2];
    c[i+3]=a[i+3]+b[i+3];
}
```

c) Unrolled code to isolate cache misses

i=0   i=4

d) Schedule for the unrolled loop (dynamic cycle count = (40 + 4 + 4 + 7) * (16/4) = 220 cycles)

```
t1=a[0];
t2=b[0];
for(i=0;i<12;i+=4){
    c[i]=t1+t2;
    t1=a[i+4];
    t2=b[i+4];
    c[i+1]=a[i+1]+b[i+1];
    c[i+2]=a[i+2]+b[i+2];
    c[i+3]=a[i+3]+b[i+3];
}
c[12]=t1+t2;
c[13]=a[13]+b[13];
c[14]=a[14]+b[14];
c[15]=a[15]+b[15];
```

e) Unrolled and shifted code to improve cache miss/hit overlaping

i=0   i=4

f) Schedule for the unrolled and shifted loop (dynamic cycle count = 40 + (3+40) * (12/4) +
+ (3+4+4+7) = 187 cycles)

Figure 5.5. Motivating example

Algorithm: MIST
Input: Memory hierarchy timing and pipelining model and Application code,
Output: Application code optimized for memory miss traffic
Begin MIST
1. Perform reuse analysis and determine the miss accesses
2. Isolate cache misses
3. Attach accurate timing to the cache hit and miss accesses.
4. Perform cache dependence analysis.
5. Perform loop shifting to reduce the dependence chains in the loop body.
6. Perform Trailblazing Percolation Scheduling (TiPS).
End MIST

Figure 5.6.   The MIST Miss Traffic optimization algorithm.

overlap the cache miss traffic (the figure does not depict the loop prologue and
epilogue). The dynamic cycle count is 187 cycles, generating a further 17.6%
performance improvement over the already optimized version from Figure 5.5
(d).

Thus, a more accurate timing model and cache behavior information allows
the compiler to more aggressively overlap the cache hit/miss operations, and
create significant performance improvement over the traditional optimistically
scheduled approach.

5.3.2   Miss Traffic Optimization Algorithm

We present in the following the Miss Traffic Optimization algorithm (called
MIST), which performs aggressive scheduling of cache miss traffic to better target the memory subsystem architecture, and generate significant performance
improvements. The MIST algorithm receives as input the application code,
along with the memory hierarchy timing model in the form of operation timings [GDN00], and the memory subsystem parallelism and pipelining model
captured as Reservation Tables [GHDN99], both generated from an EXPRES-
SION [HGG+99] description of the processor/memory system architecture.
MIST produces as output the application code optimized for cache miss traffic.

Figure 5.6 presents the Miss Traffic Optimization algorithm. The first step
performs reuse analysis and determines the array accesses resulting in misses.
The second step isolates the misses in the code through loop transformations,
and the third step attaches accurate timings to these isolated accesses. The
fourth step determines the dependences between the hits to a cache line and
the misses which fetch the data from the main memory for that cache line,
while the fifth step performs loop shifting to transform these dependences from
intra-iteration into loop-carried dependences. The last step performs Instruction Level Parallelism (ILP) extraction, to aggressively overlap the memory
operations.

Whereas the main contribution of this approach is the idea of optimizing the cache miss traffic to better utilize the memory bandwidth, the loop shifting algorithm is only an instance of such a miss transfer optimization we propose. Once the information regarding miss accesses along with the accurate timing, pipelining and parallelism models are known, other parallelizing optimizations such as software pipelining can be successfully applied.

We will use the example code from Figure 5.5 (a) to illustrate the MIST algorithm. While this example has been kept simple for clarity of the explanations (single-dimensional arrays, simple array index functions and no nested loops), our algorithm can handle multi-dimensional arrays, with complex index functions, accessed from multiple nested loops.

Step 1 performs cache behavior analysis, predicting the cache misses, while Step 2 isolates these misses in the code through loop transformations. We use a standard cache behavior analysis and miss isolation technique presented in [MLG92] and [Wol96a] [6]. For instance, the array reference a[i] from Figure 5.5 (a) results in a miss access every 4th iteration of the loop (assuming a cache line size 4), and the set of predicted cache misses is:

$$Miss\,(a[i]) = \{i \mid 0 \le i \le 15 \; and \; i \bmod 4 = 0\}$$

Figure 5.5 (c) shows the isolation of the cache misses of the array access a[i] in the original example through loop transformations. After unrolling the loop 4 times, the first access to array a in the body of the unrolled loop (a[i]) generates always a miss, while other 3 accesses (a[i+1], a[i+2], a[i+3]) generate hits.

The third step of the MIST algorithm attaches the accurate timing model for each of the array accesses, depending on whether it represents hits or misses. The timing information is received as an input by the MIST algorithm, and specifies for each operation the moment when the operands are read/written and the latency of the operation [GDN00]. Additionally, a model of the pipelining and parallelism, represented as Reservation Tables [GHDN99] is attached to each operation. Together, the timing and the pipelining/parallelism models allow the compiler to aggressively schedule these operations.

For instance, for the array access a[i] in Figure 5.5 (c) which represents cache misses, we attach the latency of 20 cycles, along with the reservation table representing the resources used during execution (e.g., fetch, decode, ld/st unit, cache controller, memory controller, main memory, etc.). The operation timing and reservation table models are detailed in [GHDN99] and [GDN00].

The miss access to a cache line is responsible for bringing the data into the cache from the main memory. Hit accesses to that cache line have to wait on

---

[6]We could also use alternative techniques such as those presented in [GMM99].

Procedure: Cache_dependence_analysis
Input: Code with isolated cache misses
Output: Cache_dependences
Begin Cache_dependence_analysis
 for each array a
  for each array access x to array a
   for each array access y to array a, different than x
    if x is a miss and y is a hit and there is group spatial locality between x and y
     Create cache dependence from x to y
End Cache_dependence_analysis

Figure 5.7. The cache dependence analysis algorithm.

Procedure: Loop_shifting
Input: Code with isolated cache misses and cache dependences
Output: Shifted code with shorter dependence chains
Begin Loop_shifting
 for each innermost loop l
  for each array access a in loop l body
   if array access a represents a miss
    if array access a has intra-iteration cache dependence to subsequent access b
     shift array access a to previous iteration of loop l
End Loop_shifting

Figure 5.8. The loop shifting algorithm.

the access which generated the miss on that line to complete, making the data available in the cache. In order to aggressively schedule such instructions, the compiler must be aware of this dependence between the hit access, and the miss access to the same cache line. In the following we will call this a "cache dependence".

The fourth step determines cache dependences between different memory accesses. The cache dependence analysis algorithm is presented in Figure 5.7. There is a cache dependence between two accesses if they refer to the same cache line, one of them represents a miss, and the other a hit. The two accesses refer to the same cache line if there is group spatial locality between them (as determined in Step 1). For instance, in the example code from Figure 5.5 (c), a[i+1] depends on a[i], since a[i] and a[i+1] refer to the same cache line, and a[i] generates a miss. The cache dependences are used in the following to guide loop shifting and parallelization, and facilitate the overlap between memory accesses which do not depend on each other.

In Step 5 of the Memory Traffic Optimization algorithm we perform loop shifting to increase the parallelism opportunity between cache miss and hit accesses. The loop shifting algorithm is presented in Figure 5.8.

Often memory accesses which address the same cache line are close together in the code (especially after performing cache optimizations such as tiling [PDN99]). As a result, the cache hits and the miss to the same line create long dependence chains, which prohibit the compiler from aggressively overlapping these memory accesses (even if the compiler would optimistically overlap them, the memory controller would insert stalls, resulting in performance penalties). Step 5 performs loop shifting to transform the cache dependences from intra-iteration dependences into loop-carried dependences. By reducing the intra-iteration dependence chains, we increase the potential parallelism in the loop body, and allow the compiler to perform more aggressive scheduling of the memory operations. In the example from Figure 5.5 (c), the miss accesses a[i] and b[i] create a cache dependence on the hits from a[i+1], b[i+1], etc.. To reduce the dependence chains in the loop body (by transforming the intra-iteration dependences into loop-carried dependences), we shift the miss accesses a[i] and b[i] to the previous iteration, as shown in Figure 5.5 (e). As a result an increased parallelism is exposed in the loop body, and the compiler can better overlap the memory operations. Of course this loop shifting technique results in increased code size but yields better performance. Therefore, for space-critical embedded applications, the designer will need to tradeoff increase in code size for improved performance.

In Step 6 we use an Instruction Level Parallelism (ILP) scheduling approach to parallelize the operations in the loops, based on the accurate timing models derived in the Step 3. While other ILP scheduling technique could be used as well to parallelize the code, we use Trailblazing Percolation Scheduling (TiPS) [NN93], a powerful ILP extraction technique which allows parallelization across basic-block boundaries.

Due to the accurate timing information, and the loop shifting which increases the potential parallelism between memory accesses, the ILP scheduling algorithm generates significantly more parallelism than in the traditional version, with optimistic timing for the memory accesses. The resulting code presents a high degree of parallelism in the memory miss traffic, efficiently utilizing the main memory bandwidth, and creating significant performance improvements.

5.3.3   Experiments

We present a set of experiments demonstrating the performance gains obtained by aggressively optimizing the memory miss traffic on a set of multimedia and DSP benchmarks. We perform the optimization in two phases: first we isolate the cache misses and attach accurate hit and miss timing to the memory accesses, to allow the scheduler to better target the memory subsystem architecture. We then further optimize the cache miss traffic, by loop shifting to reduce the intra-iteration dependence chains due to accesses to the same cache line, and allow more overlap between memory accesses. We compare both these approaches to the traditional approach, where the scheduler uses an optimistic timing to parallelize the operations, the best alternative available to the compiler, short of accurate timing and cache behavior information.

5.3.3.1   Experimental setup

In our experiments we use an architecture based on the Texas Instruments TIC6211 VLIW DSP, with a 16k direct mapped cache. The TIC6211 is an integer 8-way VLIW processor, with 2 load/store units. The latency of a hit is 2 cycles, and of a miss is 20 cycles.

The applications have been compiled using the EXPRESS [HDN00] retargetable optimizing ILP compiler, and the Trailblazing Percolation Scheduling (TiPS) [NN93] algorithm, a powerful Instruction Level Parallelism (ILP) scheduling technique. The cycle counts have been computed using our cycle-accurate structural simulator SIMPRESS [KSH+99]. The timing models have been generated using the TIMGEN Timing Generation algorithm [GDN00] and the pipelining/parallelism models in the form of Reservation Tables have been generated using the RTGEN Reservation Tables Generation algorithm [GHDN99] from an EXPRESSION description of the architecture. The timing and reservation tables generated from EXPRESSION model both the memory subsystem architecture and the processor pipeline. To clearly separate out the performance improvement obtained by the miss traffic optimization algorithm, we performed the best schedule available in the traditional version, using TiPS, and accurate timing, pipelining and parallelism information for both the processor, and the cache hit operations, while the cache miss operations are handled by the memory controller.

5.3.3.2   Results

The first column in Table 5.3 shows the benchmarks we executed (from multimedia and DSP domains). The second column represents the dynamic cycle count for the traditional approach, using an optimistic timing for the memory accesses, assuming that all accesses are cache hits, and no loop optimizations performed. The third column represents the dynamic cycle count for the first phase of our optimization, using accurate timing and cache behavior information, while the fourth column shows the corresponding percentile performance improvement over the traditional approach. The fifth column represent the dynamic cycle count for the second phase of our optimization, using both accurate timing and cache behavior information and optimized for cache miss traffic through loop shifting. The sixth column represents the percentile performance improvement over the first phase of our optimization, while the last column shows the improvement of the second phase over the base-line traditional approach.

The performance improvement due to providing the compiler with information on the memory accesses which result in hits and misses, and attaching accurate timing information for each of these (shown in columns 3 and 4) varies between 15.2% (for beam, where the cache miss isolation was conservative, optimizing only the misses in the innermost loop) and 52.8% (for gsr, where the loop body contains multiple misses which can be efficiently overlapped), resulting in an average of 34.1% gain.

The extra performance obtained by further overlapping the misses to one cache line with cache hits to different lines through loop shifting (shown in columns 5 and 6) varies between 0% (for gsr, where there are enough independent accesses in the loop body to parallelize without loop shifting), and 45.7% (for mm, where the intra-iteration dependence chain between the miss and the hits to the same cache line is significantly reduced through loop shifting), generating an average of 21.3% further improvement over the first phase of the optimization.

Benchmark	Optimistic Timing (traditional)	Accurate timing & miss info				Overall perf gain
		w/o miss traffic optimization		w/ miss traffic optimization		
		# cycles	%	# cycles	%	%
beam	134433	116629	15.2	90147	29.3	49.1
dequant	5287	3792	39.4	3089	22.7	71.1
gsr	142213	93067	52.8	93067	0	52.8
idct	31314	26816	16.7	22749	17.8	37.6
lowpass	158114	114565	38.0	108805	5.2	45.3
madd	3434	2320	48.0	1876	23.6	83.0
mm	120979	104324	15.9	71583	45.7	69.0
wavelet	2682	1830	46.5	1448	26.3	85.2
Average			**34.1**		**21.3**	**61.6**

Table 5.3. Dynamic cycle counts for the TI C6211 processor with a 16k direct mapped cache

The overall performance improvement obtained by both phases of the optimization over the traditional approach (shown in column 7), varies between 37.6% for idct (where in order to keep the degree of loop unrolling low we used conservative cache prediction and isolation information, by heuristically considering some of the hits as misses), and 85.2% (in wavelet), with an average of 61.6%. In general, when a hit is wrongly predicted as a cache miss, we schedule it earlier, generating no penalty. When a miss is wrongly predicted as a cache hit, the resulting schedule will be similar to the traditional approach.

Due to the program trasformations such as loop unrolling and shifting, the performance improvement comes at the cost of an increase in code size. In Table 5.4 we present the code size increase for the set of multimedia benchmarks.

Column 1 in Table 5.4 shows the benchmark. Column 2 shows the number of assembly lines in the original application. Column 3 shows the unrolling factor for the phase I of the optimization, while Column 4 presents the number of assembly lines for the unrolled code. Column 5 shows the percentage increase in code size for the first phase of the MIST optimization (cache miss isolation), compared to the original code size. Column 6 shows the unroll factor for the second phase of the optimization (performing further program transformations to overlap cache misses with hits to a different cache line), along with the number of memory accesses shifted to previous iterations of the loop, while the Column 7 presents the number of lines of assembly code for phase II of the miss traffic optimization. The last column shows the percentage code size increase generated by the second phase of the optimization.

Benchmark	Optimistic	Accurate timing & miss info					
	Timing	w/o miss traffic optimization			w/ miss traffic optimization		
	(traditional)	unroll factor	# asm lines	%	unroll factor/ # accesses shifted	# asm lines	%
beam	152	4	183	20.3	4/2	228	24.5
dequant	136	4	323	137.5	4/1	329	1.8
gsr	146	4	385	163.6	4/0	385	0
idct	177	4	211	19.2	4/2	258	22.2
lowpass	183	4	335	83.0	4/0	335	0
madd	67	4	103	53.7	4/2	143	38.8
mm	108	4	144	33.3	4/2	197	36.8
wavelet	128	2	186	45.3	2/1	244	31.1
Average				**69.5**			**19.4**

Table 5.4. Code size increase for the multimedia applications

For our multimedia kernels, the code size increase varies between 19% (for idct, where we limited the unrolling factor, performing conservative cache behavior analysis and miss isolation) and 163% (for gsr, where loop unrolling has a large impact on code size) for the cache miss isolation phase, and between 0% (for gsr, where there are enough cache misses in the loop body to overlap without further program transformations) and 38% (for madd, where loop shifting is needed to reduce the intra-body cache dependences) for the second phase of the algorithm.

We use a heuristic to limit the unrolling factor. As a result, some of the benchmarks produced approximate cache miss isolation (e.g., idct). Even in such imperfect miss isolation, the miss traffic optimization algorithm produced performance improvements. Code size increase is traded off against performance improvement. By permitting more unrolling and shifting, better miss isolation and timing can be obtained, resulting in higher performance improvement. For instance, idct has small code size increase and smaller performance improvement (37%), while dequant has large increase in code size, but also large performance gain (71%). Moreover, when the access patterns are uniform accross the application, memory allocation optimizations such as alligning the array to improve the overlap between cache misses and hits to a different cache line (for instance through array padding [PDN99]) can be used to replace the loop shifting transformation, thus avoiding the code size increase.

The large performance gains obtained by isolating the cache misses and attaching accurate timing, are due to the better opportunities to hide the latency of the lengthy memory operations. Furthermore, by loop shifting we reduce the dependence chains in the loop body, and create further parallelism opportunities between cache hits and misses to different cache lines. This effect is particularly large in multimedia applications showing high spatial locality in the array accesses (present in the original access pattern, or after cache optimizations such as tiling[PDN99]).

5.3.4   Related Work

We have not seen any prior work that directly addresses compiler management of cache misses using cache analysis and accurate timing to aggressively schedule the cache miss traffic. However, related work exists in 3 areas: (I) cache optimizations, which improve the cache hit ratio through program transformations and memory allocation, (II) cache behavior analysis, and cache hit prediction, and (III) memory timing extraction and exploitation.

I. Cache optimizations improving the cache hit ratio have been extensively addressed by both the embedded systems community ([CWG+98], [PDN99]) and the main-stream compiler community ([Wol96a]). Loop transformations (e.g., loop interchange, blocking) have been used to improve both the temporal and spatial locality of the memory accesses. Similarly, memory allocation techniques (e.g., array padding, tiling) have been used in tandem with the loop transformations to provide further hit ratio improvement. Harmsze et. al. [HTvM00] present an approach to allocate and lock the cache lines for stream based accesses, to reduce the interference between different streams and random CPU accesses, and improve the predictability of the run-time cache behavior.

However, often cache misses cannot be avoided due to large data sizes, or simply the presence of data in the main memory (compulsory misses). To efficiently use the available memory bandwidth and minimize the CPU stalls, it is crucial to aggressively schedule the cache misses.

Pai and Adve [PA99] present a technique to move cache misses closer together, allowing an out-of-order superscalar processor to better overlap these misses (assuming the memory system tolerates a large number of outstanding misses). Our technique is orthogonal, since we overlap cache misses with cache hits to a different cache line. That is, while they cluster the cache misses to fit into the same superscalar instruction window, we perform static scheduling to hide the latencies.

Data prefetching is another approach to improve the cache hit ratio used in general purpose processors. Software prefetching [CKP91], [GGV90], [MLG92], inserts prefetch instructions into the code, to bring data into the cache early, and improve the probability it will result in a hit. Hardware prefetching [Jou90], [PK94] uses hardware stream buffers to feed the cache with data from the main memory. On a cache miss, the prefetch buffers provide the required cache line to the cache faster than the main memory, but comparatively slower than the cache hit access. While software prefetching improves the cache hit ratio, it does not aggressively hide the latency of the remaining cache misses. Similarly, hardware prefetching improves the cache miss servicing time, but does not attempt to hide the latency of the stream buffer hits, and the stream buffer misses. We complement this work by managing the cache miss traffic to better hide the latency of such operations. Moreover, prefetching produces additional memory traffic [PK94], generating a substantial strain on the main memory bandwidth, due to redundant main memory accesses (e.g., by reading from the main memory data which is already in the cache), and often polluting the cache with useless data. While this may be acceptable for general purpose processors, due to the large power consumption of main memory accesses, in embedded applications the increased power consumption due to the extra memory accesses is often prohibitive. Furthermore, the main memory bandwidth is often limited, and the extra traffic along with the additional prefetch instructions and address calculation may generate a performance overhead which offsets the gains due to prefetching. We avoid these drawbacks by *not* inserting extra accesses to the memory subsystem, but rather scheduling the existing ones to better utilize the available memory bandwidth. We thus avoid the additional power consumption and performance overhead, by specifically targeting the main memory bandwidth, through an accurate timing, pipelining and parallelism model of the memory subsystem.

II. Cache behavior analysis predicts the number/moment of cache hits and misses, to estimate the performance of processor-memory systems [AFMW96], to guide cache optimization decisions [Wol96a], to guide compiler directed prefetching [MLG92] or more recently, to drive dynamic memory sub-system reconfiguration in reconfigurable architectures [JCMH99], [VTG+99]. We use the cache locality analysis techniques presented in [MLG92], [Wol96a] to recognize and isolate the cache misses in the compiler, and then schedule them to better hide the latency of the misses.

III. Additional related work addresses extraction and utilization of accurate memory timing in the context of interface synthesis [COB95], [Gup95], hardware synthesis [LKMM95], [PDN98], and memory-aware compilation [GDN00]. However, none of these approaches addresses intelligent management of cache miss traffic, and related optimizations. We complement this work by using the accurate timing information to better schedule the cache miss operations.

Our miss traffic optimization technique uses such an accurate timing model to manage the miss traffic and aggressively overlap the memory accesses, efficiently utilizing the memory bandwidth. Our approach works also in the presence of already optimized code (for instance cache hit ratio optimizations [PDN99]), by using accurate timing, pipelining and parallelism information to better manage the memory accesses beyond cache hits, and further improve the system performance. We first predict and isolate the cache misses in the application. We then use the exact timing and pipelining information in the form of operation timings [GDN00] and reservation tables [GHDN99] to schedule these cache misses. By allocating a higher priority to the transfers between the main memory and the cache, we ensure that the main memory bandwidth, which usually represents the bottleneck in memory-intensive applications, is effectively used, and the latencies are efficiently hidden, generating significant performance improvements.

5.4   Summary

In this chapter we presented an approach which allows the compiler to exploit efficient memory access modes, such as page-mode and burst-mode, offered by modern DRAM families. We hide the latency of the memory operations, by marrying the timing information from the memory IP module with the processor pipeline timings, to generate accurate operation timings. Moreover, we use cache behavior analysis together with accurate cache hit/miss timing information and loop transformations to better target the processor/memory system architecture.

Traditionally, the memory controller accounted for memory delays, by freezing the pipeline whenever memory operations took longer then expected. However, the memory controller has only a local view of the code, and can perform only limited optimizations (e.g., when reading an element which is already in the row buffer, it avoids the column decode step). In the presence of efficient memory accesses, it is possible to better exploit such features, and better hide the latency of the memory operations, by providing the compiler with accurate timing information for the efficient memory accesses, and allowing it to perform more global optimizations on the input behavior. Moreover, by providing the compiler with cache hit/miss prediction and accurate timing, we perform more global optimizations, to exploit the parallelism available in the memory subsystem, and better hide the latency of the miss accesses.

By better exploiting the efficient memory access modes provided by modern DRAM families and better hiding of the memory operation latencies through the accurate timing information, we obtained significant performance improvements. Furthermore, in the context of Design Space Exploration (DSE), our approach provides the system designer with the ability to evaluate and explore the use of different memory modules from the IP library, with the memory-aware compiler fully exploiting the efficient access modes of each memory component. We presented a set of experiments which separate out the gains due to the more accurate timing, from the gains due to the efficient access mode optimizations and access modes themselves. The average improvement was 23.9% over the schedule that exploits the access mode without detailed timing.

The average performance improvement for using cache hit/miss information to better model the memory access timing was 34.1% over the schedule using the traditional optimistic timing model. By further optimizing the cache miss traffic, an additional 21.3% average improvement was obtained, generating an overall 61.6% average improvement over the traditional approach.

The Memory-aware Compilation approach presented in this chapter has been implemented. We use the EXPRESSION language to describe the memory architecture, and automatically extract the timing and resource information needed by the compiler. The experimental results have been obtained using our SIMPRESS cycle accurate simulator.

Chapter 6

EXPERIMENTS

In the previous chapters we presented experiments showing the performance, cost, and power improvements obtained by our individual optimization and exploration steps. In this chapter we combine these techniques into a set of global experiments, presenting results on our overall Hardware/Software Memory Exploration methodology. We integrate our Memory Architecture Exploration, Connectivity Exploration, and Memory-Aware Compilation approaches, and present our unified experimental results on a set of embedded multimedia and scientific applications.

6.1   Experimental setup

We simulated the design alternatives using two simulator configurations: (a) for early estimation of the memory and connectivity architecture behavior we used our simulator based on the SIMPRESS [MGDN01] memory model, and SHADE [CK93], and (b) for more accurate simulation of the memory system together with the actual processor, we simulated the design alternatives using our simulator based on the SIMPRESS [MGDN01] memory model, and the Motorola PowerPC simulator. We assumed a processor based on a VLIW version of the Motorola PowerPC [Pow].

We compiled the applications using our EXPRESS [HGG+99] compiler, in two configurations: (I) a traditional compiler configuration, where the compiler optimistically schedules the memory operations, assuming the fastest memory access, and (II) the Memory-aware configuration, where the compiler uses accurate timing and resource models of the memory architectures. In order to verify the quality of the code generated by our compiler for the PowerPC processor, we compared our EXPRESS compiler for the PowerPC with the native Motorola compiler, and obtained similar results. We estimated the cost of the memory architectures (in equivalent basic gates) using figures generated by the Synopsys Design Compiler [Syn], and an SRAM cost estimation technique from [CWG+98].

6.2   Results

We used the following multimedia and scientific benchmarks: Compress and Li (from SPEC95), and Vocoder (a GSM voice encoding application).

6.2.1   The Compress Data Compression Application

Compress is a data compression application from SPEC95. The algorithm is based on a modified Lempel-Ziv method (LZW), which finds the common substrings and replaces them with a variable size code. We use the main compression and decompression routines from the *Compress* benchmark (from SPEC95) to illustrate the performance, and cost trade-offs generated by our approach. The Compression and decompression routines contain a varied set of access patterns, providing interesting memory customization opportunities.

We explore the memory architecture and compiler in three main phases: (I) Memory Module Exploration, customizing the memory architecture based on the most active access patterns in the application, (II) Connectivity exploration, evaluating and selecting a set of connectivity architectures to implement the communication required by the memory architecture, and (III) Memory-Aware Compilation, exploring the memory architecture in the presence of the full system, containing the Memory Architecture, the CPU, and the Memory Aware Compiler.

(I) Starting from the input application, we first evaluate and explore different memory architectures, by mixing and matching memory modules from a memory IP library. We guide the exploration towards the most promising designs, pruning out the non-interesting parts of the design space. Figure 6.1 presents the memory design space exploration of the access pattern customizations for the Compress application. The Compress benchmark exhibits a large variety of access patterns providing many customization opportunities. The x axis represents the cost (in number of basic gates), and the y axis represents the overall miss ratio (the miss ratio of the custom memory modules represents the number of accesses where the data is not ready when it is needed by the CPU, divided by the total number of accesses to that module).

The design points marked with a circle represent the memory architectures chosen during the exploration as promising alternatives, and fully simulated for accurate results. The design points marked only with a dot represent the exploration attempts evaluated through fast time-sampling simulation, from which the best cost/gain tradeoff is chosen at each exploration step.

The miss ratio of the Compress application varies between 22.14% for the initial cache-only architecture (for a cost of 319,634 gates), and 11.72% for a memory architecture where 3 access pattern clusters have been mapped to custom memory modules (for a cost of 335,391 gates). Based on a cost constraint (or alternatively on a performance requirement), the designer can select the memory architectures that best matches the goals of the system. The selected memory architectures are considered for further exploration.

In order to validate our space walking heuristic, and confirm that the chosen design points follow the pareto-curve-like trajectory in the design space, in Chapter 4.2 we compared the design points generated by our approach to the full simulation of the design space considering all the memory module allocations and access pattern cluster mappings for the Compress example benchmark; we found that indeed, our exploration heuristic successfully finds the most promising designs, without requiring full simulation of the entire design space.

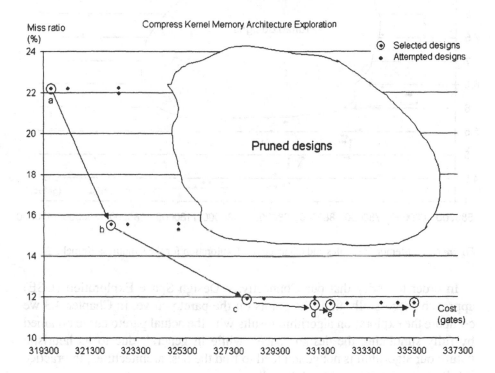

Figure 6.1.   Memory Architecture Exploration for the Compress Kernel .

(II) After selecting a set of memory modules to match the access patterns in the application, we perform connectivity exploration, by evaluating a set of connectivity modules from a connectivity IP library. For each of the selected memory configurations from the previous step, multiple connectivity architectures are possible: different standard on-chip busses (e.g., AMBA AHB, ASB, etc.) with different bit-widths, protocols, and pipelining, MUXes and dedicated connections are available, each resulting in different cost/performance tradeoffs. For each selected memory architecture, we evaluate and select the most promising connectivity architectures.

Figure 6.2 presents the connectivity exploration results. The x axis represents the cost (in number of basic gates), and the y axis represents the average memory latency, including the memory modules and connectivity latencies. We select memory and connectivity architectures following the pareto curve-like shape, and consider them for further exploration in the third phase of our approach.

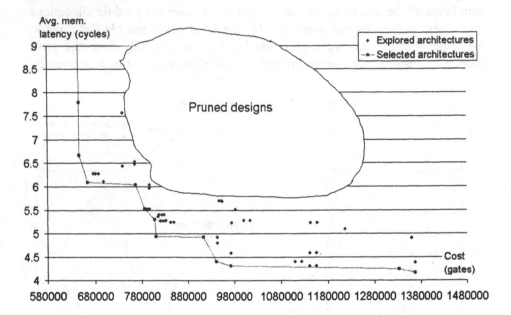

Figure 6.2. Memory Modules and Connectivity Exploration for the Compress Kernel .

In order to verify that our Connectivity Design Space Exploration (DSE) approach successfully finds the points on the pareto curve, in Chapter 4.3 we compare the exploration algorithm results with the actual pareto curve obtained by fully simulating the design space: while in general, due to its heuristic nature our algorithm is not guaranteed to find the best architectures, in practice it follows the pareto curve rather well.

(III) Once we have selected the most promising memory and connectivity architecture for the input application, we evaluate the behavior of the memory system considering the complete system, consisting of the Application, the Memory Architecture, and the Memory Aware Compiler. We use our EXPRESS Memory-Aware compiler to generate code for the selected designs, using accurate timing and resource information on the architectures. We compare the results to the code generated by the EXPRESS compiler, assuming optimistic timing information for the memory accesses. This optimistic scheduling is the best alternative available to the traditional optimizing compiler, short of an accurate timing model.

Figure 6.3 shows the exploration results using both the Memory Aware Compiler, where for each architecture we extract and use the accurate timing and resource information in the compiler, and the Traditional optimizing compiler, where we schedule the code optimistically, and rely on the memory controller to account for the longer delays. The x axis represents the cost in number of gates, while the y axis represents the cycle count. The diamond shapes (architectures a through h) represent the traditional optimizing compiler, while the squares (architectures m through u) represent the Memory Aware Compiler. Configurations a through h represent the most promising traditional compiler designs, while architectures m through u represent the most promising Memory Aware Compiler designs.

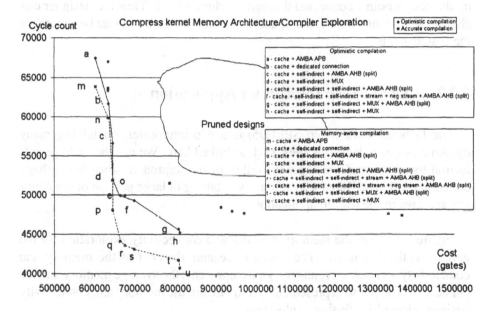

Figure 6.3. Memory Exploration for the Compress Kernel .

Starting from the traditional memory architecture and compiler configuration, our Hardware/Software Memory Exploration approach explores a much wider design space, beyond configurations that were traditionally considered. The designs a and b represent the traditional, simple cache architecture and connectivity configurations, together with an optimizing compiler which assumes an optimistic timing model for the architecture. Instead, our approach proposes architectures c through h, containing special memory modules, such as linked-list DMAs, SRAMs, stream buffers, etc., to target the access patterns in the application, and significantly improve the match between the application and the memory architecture. Next we retarget the Memory Aware Compiler to each such architecture, by providing accurate timing and resource information to the compiler, and further improve the system behavior. Designs m and n represent the traditional simple cache hierarchy with two different connectivity architectures, using our Memory Aware Compiler. By using accurate timing, resource, pipelining and parallelism information, the compiler can better hide the latency of the memory operations (designs m and n), further improving the performance (compared to the corresponding designs a and b, using the traditional optimistic compiler). Designs o through u represent novel memory architectures, and use the Memory Aware Compiler, to further expand the design space explored. Our Hardware/Software Memory Exploration approach reduces the cycle count of the Compress kernel from 67431 to 40702 cycles, for a cost increase from 603113 to 810267 gates. Moreover, a set of intermediate performance/cost tradeoffs are offered between these points (represented by the design points connected through the dotted line). Thus the designer can choose the memory, connectivity, and compiler configuration that best matches the system goals.

6.2.2   The Li Lisp Interpreter Application

The Li benchmark (from SPEC95) is a lisp interpreter, containing many dynamic memory data structures, such as linked lists. We use the garbage collection routine from Li to show the utility of our approach, since the garbage collection part is activated very often, contributing to large portion of the memory accesses in the overall application.

Figure 6.4 shows the memory modules and connectivity exploration for the garbage collection in Li. The x axis represents the cost of the memory and connectivity architectures, and the y axis represents the average memory latency per access. The points represented by a square are the memory and connectivity designs selected for further exploration.

Figure 6.5 shows the overall Hardware/Software exploration results for the garbage collection in Li. Again, starting from the traditional memory architectures, containing a simple cache, we memory architectures using stream buffers, and diverse connectivity modules. The designs d and e generate a roughly 23% performance improvement over the traditional cache only design (b). Moreover, by providing accurate timing, resource, and pipelining information to the Memory Aware Compiler, the overall system behavior is further improved. Architecture q generates a roughly 50% performance improvement over the traditional cache-only architecture using the traditional memory transparent compiler (design b).

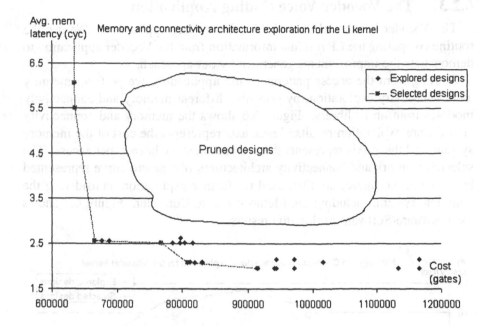

Figure 6.4.   Memory Exploration for the Li Kernel .

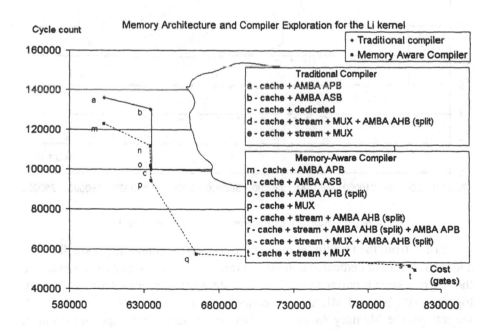

Figure 6.5.   Memory Exploration for the Li Kernel .

6.2.3   The Vocoder Voice Coding Application

The Vocoder benchmark is a GSM voice coding application. We use the routine computing the LP residual information from the Vocoder application to demonstrate the improvements generated by our approach.

Starting from the access patterns in the application, we perform memory and connectivity exploration, by selecting different memory and connectivity modules from an IP library. Figure 6.6 shows the memory and connectivity architecture exploration results. The x axis represents the cost of the memory system, and the y axis represents the average memory latency per access. The selected memory and connectivity architectures (the pareto curve represented by a line in the figure) are then used for further exploration, considering the complete system, including the Memory Aware Compiler. Figure 6.7 shows the Hardware/Software exploration results.

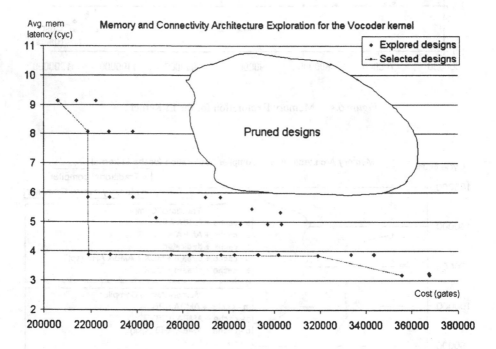

Figure 6.6.   Memory Exploration for the Vocoder Kernel .

Starting from the traditional memory architecture, containing a small cache and a bus-based or dedicated connection (architectures a, b and c), we evaluate the use of a stream buffer to implement a stream-like memory access pattern, together with bus and MUX-based connections (architectures d and e). Next, we retarget the Memory Aware Compiler, to account for the specific timings, resources, and pipelining in each such explored architectures, to further hide the latency of the memory accesses. The design points m, n and o represent

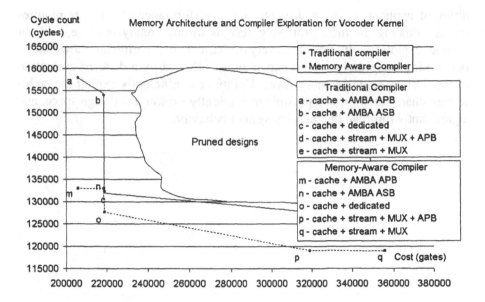

Figure 6.7. Memory Exploration for the Vocoder Kernel .

the traditional cache architectures, using the Memory Aware Compiler, while the designs p and q represent the memory architectures using a stream buffer, together with the Memory Aware Compiler.

6.3 Summary of Experiments

Clearly, by considering all three components of the embedded memory system early during the design process: the Memory Architecture, the Application, and the Memory Aware Compiler, it is possible to explore a design space well beyond the one traditionally considered.

By using specialized memory modules, such as linked-list DMAs, stream buffers, SRAMs, to match the access patterns in the application, in effect transferring some of the burden of memory accesses from software to hardware, we substantially improve the system behavior. Moreover, by providing accurate timing, resource, pipelining and parallelism information to the compiler during the exploration to better hide the latency of the memory accesses, we can generate further performance improvements. Through our combined early rapid evaluation, and detailed Compiler-in-the-loop analysis, we cover a wide range of design alternatives, allowing the designer to efficiently target the system goals. Furthermore, by pruning the non-interesting designs, we guide the search towards the most promising designs, and avoid fully simulating the design space.

The best cost/performance/power tradeoffs for different applications are provided by varied memory and connectivity architectures. Moreover, typically different applications have divergent cost, performance and power requirements. Finding the most promising designs through analysis alone, without such an exploration framework is very difficult. Our experiments show that by performing early design space exploration, followed by a detailed evaluation of the designs using Memory Aware Compiler to efficiently exploit the architecture characteristics, it is possible to efficiently explore the design space, and significantly improve the memory system behavior.

Chapter 7

CONCLUSIONS

7.1 Summary of Contributions

Recent trends, such as the ability to place more transistors on the same chip, as well as increased operating speeds have created a tremendous pressure on the design complexity and time-to-market. Moreover, in today's embedded systems memory represents a major bottleneck in terms of cost, performance, and power. With the increased processor speeds, the traditional memory gap is continuously exacerbated. More aggressive memory architectures, exploiting the needs of the access patterns present in the embedded application, together with Memory Aware compilation, considering the specific features of the memory modules (such as timings, pipelining, parallelism, special access modes, etc). are required. Moreover, early Design Space Exploration (DSE), considering all three elements of the embedded system: the Memory Architecture, the Application Access Patterns, and the Memory Aware Compiler allows for rapid evaluation of different design alternatives, allowing the designer to best match the goals of the system.

In this book we presented such an approach, where we perform Memory Architecture and Connectivity Exploration, together with Memory Aware Compilation, using memory and connectivity modules from an IP library. We use an Architectural Description Language (ADL) based approach, where the ADL specification of the processor and memory system is used to capture the architecture, and drive the automatic software toolkit generation, including the Memory Aware Compiler, and Simulator. By extracting, analyzing and clustering the Access Patterns in the application we select specialized memory modules such as stream buffers, SRAMs, Linked-list DMAs, as well as traditional modules, such as caches, and DRAMs from a memory IP library, exploring a design space beyond the one traditionally considered. Furthermore, we provide explicit memory timing, pipelining and parallelism information to a Memory-Aware Compiler, performing global optimizations to better hide the latency of the lengthy memory operations.

By using our two-phased approach, starting with an early architecture exploration, to guide the search towards the most promising designs, followed by our Compiler-in-the-loop detailed evaluation of the architectures we are able to efficiently explore the memory design alternatives without fully simulating the design space.

We presented a set of experiments on a set of large real-life multimedia and scientific applications, showing significant performance improvements, for varied cost and power footprints.

7.2   Future Directions

The work presented in this book can be extended in different directions in the future.

With the advance of memory technology, new modules exhibiting diverse features will continuously appear. While our approach covers features such as any level of pipelining and parallelism, it is likely that novel idiosyncrasies of such modules will require novel optimizations. In order to fully exploit the capabilities of such memory modules, compiler techniques to account for these features will be needed.

The exploration approach we presented is targeted towards the data memory accesses and data memory architecture. Similar techniques can be used for the instruction memory accesses, by employing memory modules such as instruction caches, on-chip ROMs, and more novel modules, such as trace caches.

The exploration approach we presented uses a heuristic to prune the design space and guide the search towards the most promising designs. Providing different space walking techniques, such as multi-criteria optimizations, or Integer Linear Programming (ILP) based techniques can further improve the optimality of the system.

In this book we focused mainly on the software part of a Hw/Sw Embedded System. Considering the interaction between the hardware on-chip ASICs and the CPU cores, using different memory sharing and communication models is a very interesting direction of research. Another possible extension is using multicore-based System-on-chip, containing multiple CPU cores, ASICs, and memory IP modules, further raising the level of abstraction.

References

[AFMW96]   M. Alt, C. Ferdinand, F. Martin, and R. Wilhelm. "Cache behavior prediction by abstract interpretation,". In *SAS*, 1996.

[ARM]   ARM   AMBA   Bus   Specification. *http://www.arm.com/armwww.ns4/html/AMBA?OpenDocument*.

[BCM95]   F. Balasa, F. Cathoor, and H. De Man. "Background memory area esti mation for multidimensional signal processing systems,,". *IEEE Trans. on VLSI Systems*, 3(2), 1995.

[BENP93]   U. Banerjee, R. Eigenmann, A. Nicolau, and D. A. Padua. "Automatic program parallelization,,". *IEEE*, 81(2), 1993.

[BG95]   S. Bakshi and D. Gajski. "A memory selection algorithm for high-performance pipelines,". In *EURO-DAC*, 1995.

[BV01]   F. Baboescu and G. Varghese. "Scalable packet classification,". In *SIGCOMM*, 2001.

[cC94]   Tzi cker Chiueh. "Sunder: A programmable hardware prefetch architecture for numerical loops,". In *Conference on High Performance Networking and Computing*, 1994.

[CF91]   R. Cytron and J. Ferrante. "Efficiently computing static single assignment form and the control dependence graph,". In *TOPLAS*, 1991.

[CK93]   R. Cmelik and D. Keppel. "Shade: A fast instruction set simulator for execution profiling,". Technical report, SUN MICROSYSTEMS, 1993.

[CKP91]   D. Callahan, K. Kennedy, and A. Porterfield. "Software prefetching,". In *ASPLOS*, 1991.

[COB95]   P. Chou, R. Ortega, and G. Borriello. "Interface co-synthesis techniques for embedded systems,". In *ICCAD*, 1995.

[CWG+98]   F. Catthoor, S. Wuytack, E. De Greef, F. Balasa, L. Nachtergaele, and A. Vandecappelle. *Custom Memory Management Methodology*. Kluwer, 1998.

[CZY⁺99]   H-M. Chen, H. Zhou, F. Young, D. Wong, H. Yang, and N. Sherwani. "Integrated floorplanning and interconnect planning,". In *ICCAD*, 1999.

[DCD97]   E. DeGreef, F. Catthoor, and H. DeMan. "Array placement for storage size reduction in embedded multimedia systems,". In *Intn'l. Conf. Application-specific Systems, Architectures and Processors*, 1997.

[DIJ95]   J-M. Daveau, T. Ben Ismail, and A. Jerraya. "Synthesis of system-level communication by an allocation-based approach,". In *ISSS*, 1995.

[DM01]   Y. Deng and W. Maly. "Interconnect characteristics of 2.5-d system integration scheme,". In *ISPD*, 2001.

[Emb]   The Embedded Microprocessor Benchmarking Consortium. *http://www.eembc.org*.

[Fre93]   M. Freericks. "The nML machine description formalism,". Technical Report TR SM-IMP/DIST/08, TU Berlin CS Dept., 1993.

[FSS99]   W. Forniciari, D. Sciuto, and C. Silvano. "Power estimation for architectural exploration of hw/sw communication on system-level busses,". In *CODES*, 1999.

[Gai]   Gaisler Research. *www.gaisler.com/leon.html*.

[GBD98]   P. Grun, F. Balasa, and N. Dutt. "Memory size estimation for multimedia applications,". In *Proc. CODES/CACHE*, 1998.

[GDN99]   P. Grun, N. Dutt, and A. Nicolau. "Extracting accurate timing information to support memory-aware compilation,". Technical report, University of California, Irvine, 1999.

[GDN00]   P. Grun, N. Dutt, and A. Nicolau. "Memory aware compilation through accurate timing extraction,". In *DAC*, Los Angeles, 2000.

[GDN01a]   P. Grun, N. Dutt, and A. Nicolau. "Access pattern based local memory customization for low power embedded systems.,". In *DATE*, 2001.

[GDN01b]   P. Grun, N. Dutt, and A. Nicolau. "APEX: Access pattern based memory architecture exploration,". In *ISSS*, 2001.

[GDN01c]   P. Grun, N. Dutt, and A. Nicolau. "Exploring memory architecture through access pattern analysis and clustering,". Technical report, University of California, Irvine, 2001.

[GDN02]   P. Grun, N. Dutt, and A. Nicolau. "Memory system connectivity exploration,". In *DATE*, 2002.

[GGV90]   E. Gornish, E. Granston, and A. Veidenbaum. "Compiler-directed data prefetching in multiprocessors with memory hierarchies,". In *ICS*, 1990.

[GHDN99]   P. Grun, A. Halambi, N. Dutt, and A. Nicolau. "RTGEN: An algorithm for automatic generation of reservation tables from architectural descriptions,". In *ISSS*, San Jose, CA, 1999.

[GMM99]   S. Ghosh, M. Martonosi, and S. Malik. "Cache miss equations: A compiler framework for analyzing and tuning memory behavior,". *TOPLAS*, 21(4), 1999.

[Gup95]   R. Gupta. *Co-Synthesis of Hardware and Software for Digital Embedded Systems*. Kluwer, 1995.

[GV98]   Tony Givargis and Frank Vahid. "Interface exploration for reduced power in core-based systems,". In *ISSS*, 1998.

[GVNG94]   D. Gajski, F. Vahid, S. Narayan, and G. Gong. *Specification and Design of Embedded Systems*. Prentice Hall, 1994.

[Gyl94]   J. Gyllenhaal. "A machine description language for compilation,". Master's thesis, Dept. of EE, UIUC,IL., 1994.

[HD97]   G. Hadjiyiannis and S. Devadas. "ISDL: An instruction set description language for retargetability,". In *Proc. DAC*, 1997.

[HDN00]   A. Halambi, N. Dutt, and A. Nicolau. "Customizing software toolkits for embedded systems-on-chip,". In *DIPES*, 2000.

[HGG⁺99]   A. Halambi, P. Grun, V. Ganesh, A. Khare, N. Dutt, and A. Nicolau. "EXPRESSION: A language for architecture exploration through compiler/simulator retargetability,". In *Proc. DATE*, March 1999.

[HHN94]   J. Hummel, L Hendren, and A Nicolau. "A language for conveying the aliasing properties of dynamic, pointer-based data structures,". In *Proceedings of the 8th International Parallel Processing Symposium*, 1994.

[hot]   *HotChips conference, '97 - '99*.

[HP90]   J. Hennessy and D. Patterson. *Computer Architecture: A quantitative approach*. Morgan Kaufmann Publishers Inc, San Mateo, CA, 1990.

[HTvM00]   F. Harmsze, A. Timmer, and J. van Meerbergen. "Memory arbitration and cache management in stream-based systems,". In *DATE*, 2000.

[HWO97]   P. Hicks, M. Walnock, and R.M. Owens. "Analysis of power consumption in memory hierarchies,". In *ISPLED*, 1997.

[IBM]   IBM Microelectronics, Data Sheets for Synchronous DRAM IBM0316409C. *www.chips.ibm.com/products/memory/08J3348/*.

[IDTS00]   S. Iyer, A. Desai, A. Tambe, and A. Shelat. "A classifier for next genration content and policy based switches,". In *HotChips*, 2000.

[JCMH99]   T. Johnson, D. Connors, M. Merten, and W. Hwu. "Run-time cache bypassing,". *Transactions on Computers*, 48(12), 1999.

[Jou90]   N. Jouppi. "Improving direct-mapped cache performance by the addition of a small fully-associative cache and prefetch buffers,". In *ISCA*, 1990.

[KHW91]   R. Kessler, M. Hill, and D. Wood. "A comparison of trace-sampling techniques for multi-megabyte caches,". Technical report, University of Wisconsin, 1991.

[KP87]   F. Kurdahi and A. Parker. "Real: A program for register allocation,". In *DAC*, 1987.

[KPDN98]   A. Khare, P. R. Panda, N. D. Dutt, and A. Nicolau. "High level synthesis with synchronous and rambus drams,". In *SASIMI*, Japan, 1998.

[KSH+99]   A. Khare, N. Savoiu, A. Halambi, P. Grun, N. Dutt, and A. Nicolau. "V-SAT: A visual specification and analysis tool for system-on-chip exploration,". In *Proc. EUROMICRO*, October 1999.

[Kul01]   C. Kulkarni. *Cache optimization for Multimedia Applications*. PhD thesis, IMEC, 2001.

[LKMM95]   T. Ly, D. Knapp, R. Miller, and D. MacMillen. "Scheduling using behavioral templates,". In *DAC*, San Francisco, 1995.

[LM98]   R. Leupers and P. Marwedel. "Retargetable code generation based on structural processor descriptions,". *Design Automation for Embedded Systems*, 3(1), 1998.

[LMVvdW93]   P.E.R Lippens, J.L.Van Meerbergen, W.F.J. Verhaegh, and A. van der Wef. "Allocation of multiport memories for hierarchical data streams,". In *ICCAD*, 1993.

[LRLD00]   K. Lahiri, A Raghunatan, G. Lakshminarayana, and S. Dey. "Communication architecture tuners: A methodology for hte deisng of high-performance communication architectures for systems-on-chip,". In *DAC*, 2000.

[LS94]   D. Liu and C. Svenson. "Power consumption estimation in cmos vlsi chips,". *IEEE J. of Solid stage Circ.*, 29(6), 1994.

[MDK01]   Seapahn Maguerdichian, Milenko Drinic, and Darko Kirovski. "Latency-driven design of multi-purpose systems-on-chip,". In *DAC*, 2001.

[MGDN01]   P. Mishra, P. Grun, N. Dutt, and A. Nicolau. "Processor-memory co-explotation driven by a memory-aware architecture description language,". In *International Conference on VLSI Design*, Bangalore, India, 2001.

[MLG92]   T. Mowry, M. Lam, and A Gupta. "Design and evaluation of a compiler algorithm for prefetching,". In *ASPLOS*, 1992.

[mpr]   *Microprocessor Report*.

[NG94]   Sanjiv Narayan and Daniel D. Gajski. "Protocol generation for communication channels,". In *DAC*, 1994.

[NN93]   A. Nicolau and S. Novack. "Trailblazing: A hierarchical approach to percolation scheduling,". In *ICPP*, St. Charles, IL, 1993.

[PA99]   V. Pai and S. Adve. "Code transformations to improve memory parallelism,". In *MICRO*, 1999.

[PDN97]   P. Panda, N. Dutt, and A. Nicolau. "Architectural exploration and optimization of local memory in embedded systems,". In *ISSS*, 1997.

[PDN98]   P. R. Panda, N. D. Dutt, and A. Nicolau. "Exploiting off-chip memory access modes in high-level synthesis,". In *IEEE Transactions on CAD*, February 1998.

[PDN99]   P. Panda, N. Dutt, and N. Nicolau. *Memory Issues in Embedded Systems-on-Chip*. Kluwer, 1999.

[PEK⁺95a]   A. Purakayashta, C. Ellis, D. Kotz, N. Nieuwejaar, and M. Best. "Characterizing parallel file access patterns on a large-scale multiprocessor,". In *IPPS*, 1995.

[PEK⁺95b]   A. Purakayastha, C. Ellis, D Kotz, N. Nieuwejaar, and M. Best. "Characterizing parallel file access patterns on a large-scale multiprocessor,". In *IPPS*, 1995.

[PGG⁺95]   R. Patterson, G. Gibson, E. Ginting, D. Stodolsky, and J. Zelenka. "Informed prefeching and caching,". In *SIGOPS*, 1995.

[PK94]   S. Palacharla and R. Kessler. "Evaluating stream buffers as a secondary cache replacement,". In *ISCA*, 1994.

[Pow]   The   PowerPC   Architecture,   Motorola   Inc. *http://www.mot.com/SPS/RISC/smartnetworks/arch/powerpc/index.htm*.

[Pri96]   Betty Prince. *High Performance Memories, New Architecture DRAMs and SRAMs evolution and function*. Wiley, West Sussex, 1996.

[Prz97]   S. Przybylski. "Sorting out the new DRAMs,". In *Hot Chips Tutorial*, Stanford, CA, 1997.

[PS97]   F. Catthoor G. de Jong P. Slock, S. Wuytack. "Fast and extensive system-level memory exploration for atm applications,". In *ISSS*, 1997.

[PTVF92]   W. H. Press, S. A. Teukolsky, W. T. Vetterling, and B. P. Flannery. *Numerical Recipes in C: The Art of Scientific Computing*. Cambridge University Press, 1992.

[PUSS]   I. Parsons, R. Unrau, J. Schaeffer, and D. Szafron. "Pi/ot: Parallel i/o templates,". In *Parallel Computing, Vol. 23, No. 4-5, pp. 543-570, May 1997*.

[RGC94]   L. Ramachandran, D. Gajski, and V. Chaiyakul. "An algorithm for array variable clustering,". In *EDTC*, 1994.

[RMS98]   A. Roth, A. Moshovos, and G. Sohi. "Dependence based prefetching for linked data structures,". In *ASPLOS*, 1998.

[Sem98]   Semiconductor Industry Association. *National technology roadmap for semiconductors: Technology needs*, 1998.

[ST97]   H. Schmit and D. E. Thomas. "Synthesis of application-specific memory designs,". *IEEE Trans. on VLSI Systems*, 5(1), 1997.

[Syn]   Synopsys Design Compiler. *www.synopsys.com*.

[Tex]   Texas Instruments. *TMS320C6201 CPU and Instruction Set Reference Guide.*

[Tri97]   Trimaran Release: http://www.trimaran.org. *The MDES User Manual,* 1997.

[TSKS87]   C. Thacker, L. Steward, and Jr. K. Satterthwaite. "Firefly: A multiprocessor workstation,". Technical Report 23, DEC SRC Report, 1987.

[VGG94]   F. Vahid, D. Gajski, and J. Gong. "A binary constraint search algorithm for minimizing hardware during hardware/software partitionig,". In *EDAC,* 1994.

[VKI⁺00]   N. Vijaykrishnan, M. Kandemir, M. J. Irwin, H. S. Kim, and W. Ye. "Energy-driven integrated hardware-software optimizations using simplepower,". In *ISCA,* 2000.

[vSFCM93]   M. van Swaaij, F. Franssen, F. Catthoor, and H. De Man. "High-level modeling of data and control flow for signal processing systems,". *Design Methodologies for VLSI DSP Architectures and Applications,* 1993.

[VSR94]   I. Verbauwhede, C. Scheers, and J. Rabaey. "Memory estimation for high-level synthesis,". In *DAC,* 1994.

[VTG⁺99]   A. Veidenbaum, W. Tang, R. Gupta, A. Nicolau, and X. Ji. "Adapting cache line size to application behavior,". In *ICS,* 1999.

[WCdJ⁺96]   S. Wuytack, F. Catthoor, G. de Jong, B. Lin, and H. De Man. "Flow graph balancing for minimizing the required memory bandwith,". In *ISSS,* La Jolla, CA, 1996.

[WJ96]   S. J. E. Wilton and N. P Jouppi. "Cacti:an enhanced cache access and cycle time model,". *IEEE Journal of Solid-State Circuits,* 31(5), 1996.

[Wol96a]   W. Wolf. "Object oriented cosynthesis of distributed embedded systems,". *TODAES,* 1(3), 1996.

[Wol96b]   M. Wolfe. *High Performance Compilers for Parallel Computing.* Addison-Wesley, 1996.

[ZM99]   Y. Zhao and S. Malik. "Exact memory estimation for array computations without loop unrolling,". In *DAC,* 1999.

[ZMSM94]   V. Zivojnovic, J. Martinez, C. Schlager, and H. Meyr. "Dspstone: A dsp-oriented benchmarking methodology,". In *ICSPAT,* 1994.

Index